VOLVO
PV444 & PV544
1945-1965

Compiled by
R.M. Clarke

ISBN 1 85520 1720

Booklands Books Ltd.
PO Box 146, Cobham, KT11 1LG
Surrey, England

BROOKLANDS BOOKS

BROOKLANDS ROAD TEST SERIES
AC Ace & Aceca 1953-1983
Alfa Romeo Alfasud 1972-1984
Alfa Romeo Alfetta Coupes GT. GTV. GTV6 1974-1987
Alfa Romeo Giulia Berlinas 1962-1976
Alfa Romeo Giulia Coupes Gold Portfolio 1963-1976
Alfa Romeo Giulia Coupes 1963-1976
Alfa Romeo Giulietta Gold Portfolio 1954-1965
Alfa Romeo Spider Gold Portfolio 1966-1991
Alfa Romeo Spider 1966-1990
Allard Gold Portfolio 1937-1959
Alvis Gold Portfolio 1919-1967
American Motors Muscle Cars 1966-1970
Armstrong Siddeley Gold Portfolio 1945-1960
Aston Martin Gold Portfolio 1972-1985
Austin Seven 1922-1982
Austin A30 & A35 1951-1962
Austin Healey 100 & 100/6 Gold Portfolio 1952-1959
Austin Healey 3000 Gold Portfolio 1959-1967
Austin Healey Sprite 1958-1971
Avanti 1962-1990
BMW Six Cylinder Coupes 1969-1975
BMW 1600 Col. 1 1966-1981
BMW 2002 1968-1976
BMW 316, 318, 320 Gold Portfolio 1975-1990
BMW 320, 323, 325 Gold Portfolio 1977-1990
Buick Automobiles 1947-1960
Buick Muscle Cars 1965-1970
Buick Riviera 1963-1978
Cadillac Automobiles 1949-1959
Cadillac Automobiles 1960-1969
Cadillac Eldorado 1967-1978
Chevrolet Camaro SS & Z28 1966-1973
Chevrolet Camaro & Z-28 1973-1981
High Performance Camaros 1982-1988
Camaro Muscle Portfolio 1967-1973
Chevrolet 1955-1957
Chevrolet Corvair 1959-1969
Chevrolet Impala & SS 1958-1971
Chevrolet Muscle Cars 1966-1971
Chevelle and SS 1964-1972
Chevy Blazer 1969-1981
Chevy EL Camino & SS 1959-1987
Chevy II Nova & SS 1962-1973
Chrysler 300 Gold Portfolio 1955-1970
Citroen Traction Avant Gold Portfolio 1934-1957
Citroen DS & ID 1955-1975
Chevrolet Corvette Gold Portfolio 1968-1977
High Performance Corvettes 1983-1989
Daimler SP250 Sport & V-8250 Saloon Gold Portfolio 1959-1969
Datsun 240Z 1970-1973
Datsun 280Z & ZX 1975-1983
De Tomaso Collection No.1 1962-1981
Dodge Charger 1966-1974
Dodge Muscle Cars 1967-1970
Excalibur Collection No.1 1952-1981
Facel Vega 1954-1964
Ferrari Cars 1946-1956
Ferrari Dino 1965-1974
Ferrari Dino 308 1974-1979
Ferrari 308 & Mondial 1980-1984
Ferrari Collection No.1 1960-1970
Fiat-Bertone X1/9 1973-1988
Fiat Pininfarina 124 + 2000 Spider 1968-1985
Ford Automobiles 1949-1959
Ford Aanchero 1957-1959
Ford Bronco 1966-1977
Ford Bronco 1978-1988
Ford Consul. Zephyr Zodiac MkI & II 1950-1962
Ford Cortina 1600E & GT 1967-1970
Ford Fairlane 1955-1970
Ford Falcon 1960-1970
Ford GT40 Gold Portfolio 1964-1987
Ford Zephyr Zodiac Executive MkIII & MkIV 1962-1971
High Performance Capris Gold Portfolio 1969-1987
High Performance Escorts Mk1 1968-1974
High Performance Escorts Mk II 1975-1980
High Performance Escorts 1980-1985
High Performance Fiestas 1985-1990
High Performance Fiestas 1979-1991
High Performance Mustangs 1982-1988
Holden 1948-1962
Honda CRX 1983-1987
Hudson & Railton 1936-1940
Jaguar and SS Gold Portfolio 1931-1951
Jaguar XK120 XK140 XK150 Gold Portfolio 1948-1960
Jaguar MkVII VIII IX X 420 Gold Portfolio 1950-1970
Jaguar Cars 1961-1964
Jaguar Mk2 1959-1969
Jaguar E-Type Gold Portfolio 1961-1971
Jaguar E-Type 1966-1971
Jaguar E-Type V-12 1971-1975
Jaguar XJ12 XJ5.3 V12 Gold Portfolio 1972-1990
Jaguar XJ6 Series II 1973-1979
Jaguar XJ6 Series III 1979-1986
Jaguar XJS Gold Portfolio 1975-1990
Jeep CJ5 & CJ6 1960-1976
Jeep CJ5 & CJ7 1976-1986
Jensen Cars 1946-1967
Jensen Cars 1967-1979
Jensen Interceptor Gold Portfolio 1966-1986
Jensen Healey 1972-1976
Lamborghini Cars 1964-1970
Lamborghini Countach & Urraco 1974-1980
Lamborghini Countach & Jalpa 1980-1985
Lancia Fulvia Gold Portfolio 1963-1976
Lancia Stratos 1972-1985
Land Rover Series I 1948-1958
Land Rover Series II & IIa 1958-1971
Land Rover Series III 1971-1985
Land Rover 90 & 110 1983-1989
Lincoln Gold Portfolio 1949-1960
Lincoln Continental 1961-1969
Lincoln Continental 1969-1976
Lotus and Caterham Seven Gold Portfolio 1957-1989
Lotus Cortina Gold Portfolio 1963-1970
Lotus Elan Gold Portfolio 1962-1974
Lotus Elan Collection No.2 1963-1972
Lotus Elite 1957-1964
Lotus Elite & Eclat 1974-1982
Lotus Turbo Esprit 1980-1986
Lotus Europa Gold Portfolio 1966-1975
Marcos Cars 1960-1988
Maserati 1965-1970
Maserati 1970-1975
Mazda RX-7 Collection No.1 1978-1981
Mercedes 190 & 300SL 1954-1963
Mercedes 230/250/280SL 1963-1971

Mercedes Benz SLs & SLCs Gold Portfolio 1971-1989
Mercedes Benz Cars 1949-1954
Mercedes Benz Cars 1954-1957
Mercedes Benz Cars 1957-1961
Mercedes Benz Competition Cars 1950-1957
Mercury Muscle Cars 1966-1971
Metropolitan 1954-1962
MG TC 1945-1949
MG TD 1949-1953
MG TF 1953-1955
MG Cars 1959-1962
MGA & Twin Cam Gold Portfolio 1955-1962
MGB MGC & V8 Gold Portfolio 1962-1980
MGB Roadsters 1962-1980
MGB GT 1965-1980
MG Midget 1961-1980
Mini Cooper Gold Portfolio 1961-1971
Mini Moke 1964-1989
Mini Muscle Cars 1961-1979
Mopar Muscle Cars 1964-1967
Morgan Three-Wheeler Gold Portfolio 1910-1952
Morgan Cars 1960-1970
Morgan Cars Gold Portfolio 1968-1989
Morris Minor Collection No.1
Mustang Muscle Cars 1967-1971
Oldsmobile Automobiles 1955-1963
Old's Cutlass & 4-4-2 1964-1972
Oldsmobile Muscle Cars 1964-1971
Oldsmobile Toronado 1966-1978
Opel GT 1968-1973
Packard Gold Portfolio 1946-1958
Pantera Gold Portfolio 1970-1989
Panther Gold Portfolio 1972-1990
Plymouth Barracuda 1964-1974
Plymouth Muscle Cars 1967-1971
Pontiac Tempest & GTO 1961-1965
Pontiac Firebird and Trans-Am 1973-1981
High Performance Firebirds 1982-1988
Pontiac Fiero 1984-1988
Pontiac Muscle Cars 1966-1972
Porsche 356 1952-1965
Porsche Cars in the 60's
Porsche Cars 1960-1964
Porsche Cars 1964-1968
Porsche Cars 1968-1972
Porsche Cars 1972-1975
Porsche 911 1965-1969
Porsche 911 1970-1972
Porsche 911 1973-1977
Porsche 911 Carrera 1973-1977
Porsche 911 Turbo 1975-1984
Porsche 911 SC 1978-1983
Porsche 914 Gold Portfolio 1969-1976
Porsche 924 Gold Portfolio 1975-1988
Porsche 928 1977-1989
Range Rover Gold Porfolio 1970-1992
Reliant Scimitar 1964-1986
Riley 11/2 & 21/2 Litre Gold Portfolio 1945-1955
Rolls Royce Silver Cloud Gold Portfolio 1955-1965
Rolls Royce Silver Shadow 1965-1981
Rover P4 1949-1959
Rover P4 1955-1964
Rover 3 & 3.5 Litre Gold Portfolio 1958-1973
Rover 2000 + 2200 1963-1977
Rover 3500 1968-1977
Rover 3500 & Vitesse 1976-1986
Saab Sonett Collection No.1 1966-1974
Saab Turbo 1976-1983
Shelby Mustang Muscle Portfolio 1965-1970
Stubebaker Gold Portfolio 1947-1966
Stubebaker Hawks & Larks 1956-1963
Sunbeam Tiger & Alpine Gold Portfolio 1959-1967
Thunderbird 1955-1957
Thunderbird 1958-1963
Thunderbird 1964-1976
Toyota Land Cruiser 1956-1984
Toyota MR2 1984-1988
Triumph 2000. 2.5. 2500 1963-1977
Triumph GT6 1966-1974
Triumph Spitfire Gold Portfolio 1962-1980
Triumph Stag 1970-1980
Triumph TR2 & TR3 1952-60
Triumph TR4-TR5-TR250 1961-1968
Triumph TR6 Gold Portfolio 1969-1976
Triumph TR7 & TR8 1975-1982
Triumph Herald 1959-1971
Triumph Vitesse 1962-1971
TVR Gold Portfolio 1959-1990
Valiant 1960-1962
VW Beetle Collection No.1 1970-1982
VW Golf GTi 1976-1986
VW Karmann Ghia 1955-1982
VW Kubelwagen 1940-1975
VW Scirocco 1974-1981
VW Bus. Camper. Van 1954-1967
VW Bus. Camper. Van 1968-1979
VW Bus. Camper. Van 1979-1989
Volvo 120 1956-1970
Volvo 1800 Gold Portfolio 1960-1973

BROOKLANDS ROAD & TRACK SERIES
Road & Track on Alfa Romeo 1949-1963
Road & Track on Alfa Romeo 1964-1970
Road & Track on Alfa Romeo 1971-1976
Road & Track on Alfa Romeo 1977-1989
Road & Track on Aston Martin 1962-1990
Road & Track on Auburn Cord and Duesenburg 1952-1984
Road & Track on Audi & Auto Union 1952-1980
Road & Track on Audi 1980-1986
Road & Track on Austin Healey 1953-1970
Road & Track on BMW Cars 1966-1974
Road & Track on BMW Cars 1975-1978
Road & Track on BMW Cars 1979-1983
Road & Track on Cobra, Shelby & GT40 1962-1983
Road & Track on Corvette 1953-1967
Road & Track on Corvette 1968-1982
Road & Track on Corvette 1982-1986
Road & Track on Corvette 1986-1990
Road & Track on Datsun Z 1970-1983
Road & Track on Ferrari 1950-1968
Road & Track on Ferrari 1968-1974
Road & Track on Ferrari 1975-1981
Road & Track on Ferrari 1981-1984
Road & Track on Ferrari 1984-1988
Road & Track on Fiat Sports Cars 1968-1987
Road & Track on Jaguar 1950-1960
Road & Track on Jaguar 1961-1968

Road & Track on Jaguar 1968-1974
Road & Track on Jaguar 1974-1982
Road & Track on Jaguar 1983-1989
Road & Track on Lamborghini 1964-1985
Road & Track on Lotus 1972-1981
Road & Track on Maserati 1952-1974
Road & Track on Maserati 1975-1983
Road & Track on Mazda RX7 1978-1986
Road & Track on Mazda RX7 & MX5 Miata 1986-1991
Road & Track on Mercedes 1952-1962
Road & Track on Mercedes 1963-1970
Road & Track on Mercedes 1971-1979
Road & Track on Mercedes 1980-1987
Road & Track on MG Sports Cars 1949-1961
Road & Track on MG Sports Cars 1962-1980
Road & Track on Mustang 1964-1977
Road & Track on Nissan 300-ZX & GT Turbo 1984-1989
Road & Track on Peugeot 1955-1986
Road & Track on Pontiac 1960-1983
Road & Track on Porsche 1951-1967
Road & Track on Porsche 1968-1971
Road & Track on Porsche 1972-1975
Road & Track on Porsche 1975-1978
Road & Track on Porsche 1979-1982
Road & Track on Porsche 1982-1985
Road & Track on Porsche 1985-1988
Road & Track on Rolls Royce & B'ley 1950-1965
Road & Track on Rolls Royce & B'ley 1966-1984
Road & Track on Saab 1955-1985
Road & Track on Toyota Sports & GT Cars 1966-1984
Road & Track on Triumph Sports Cars 1953-1967
Road & Track on Triumph Sports Cars 1967-1974
Road & Track on Triumph Sports Cars 1974-1982
Road & Track on Volkswagen 1951-1968
Road & Track on Volkswagen 1968-1978
Road & Track on Volkswagen 1978-1985
Road & Track on Volvo 1957-1974
Road & Track on Volvo 1975-1985
Road & Track - Henry Manney at Large and Abroad

BROOKLANDS CAR AND DRIVER SERIES
Car and Driver on BMW 1955-1977
Car and Driver on BMW 1974-1987
Car and Driver on Cobra, Shelby & Ford GT 40 1963-1984
Car and Driver on Corvette 1956-1967
Car and Driver on Corvette 1968-1977
Car and Driver on Corvette 1978-1982
Car and Driver on Corvette 1983-1988
Car and Driver on Datsun Z 1600 & 2000 1966-1984
Car and Driver on Ferrari 1955-1962
Car and Driver on Ferrari 1963-1975
Car and Driver on Ferrari 1976-1983
Car and Driver on Mopar 1956-1967
Car and Driver on Mopar 1968-1975
Car and Driver on Mustang 1964-1972
Car and Driver on Pontiac 1961-1975
Car and Driver on Porsche 1955-1962
Car and Driver on Porsche 1963-1970
Car and Driver on Porsche 1970-1976
Car and Driver on Porsche 1977-1981
Car and Driver on Porsche 1982-1986
Car and Driver on Saab 1956-1985
Car and Driver on Volvo 1955-1986

BROOKLANDS PRACTICAL CLASSICS SERIES
PC on Austin A40 Restoration
PC on Land Rover Restoration
PC on Metalworking in Restoration
PC on Midget/Sprite Restoration
PC on Mini Cooper Restoration
PC on MGB Restoration
PC on Morris Minor Restoration
PC on Sunbeam Rapier Restoration
PC on Triumph Herald/Vitesse
PC on Triumph Spitfire Restoration
PC on VW Beetle Restoration
PC on 1930s Car Restoration

BROOKLANDS HOT ROD 'MUSCLECAR & HI-PO ENGINE SERIES
Chevy 265 & 283
Chevy 302 & 327
Chevy 348 & 409
Chevy 350 & 400
Chevy 396 & 427
Chevy 454 thru 512
Chrysler Hemi
Chrysler 273, 318, 340 & 360
Chrysler 361, 383, 400, 413, 426, 440
Ford 289, 302, Boss 302 & 351W
Ford 351C & Boss 351
Ford Big Block

BROOKLANDS MILITARY VEHICLES SERIES
Allied Mil. Vehicles No.1 1942-1945
Allied Mil. Vehicles No.2 1941-1946
Off Road Jeeps 1944-1971
Complete WW2 Military Jeep Manual
US Military Vehicles 1941-1945
US Army Military Vehicles WW2-TM9-2800

BROOKLANDS HOT ROD RESTORATION SERIES
Auto Restoration Tips & Techniques
Basic Bodywork Tips & Techniques
Basic Painting Tips & Techniques
Camaro Restoration Tips & Techniques
Chevrolet High Performance Tips & Techniques
Chevy-GMC Pickup Repair
Custom Painting Tips & Techniques
Engine Swapping Tips & Techniques
Ford Pickup Repair
How to Build a Street Rod
Mustang Restoration Tips & Techniques
Performance Tuning - Chevrolets of the '60s
Performance Tuning - Ford of the '60s
Performance Tuning - Mopars of the '60s
Performance Tuning - Pontiacs of the '60s

CONTENTS

Page	Title	Source	Date
5	A Day Out in a Volvo	*Motor*	July 23 1947
8	A Post War Small Car from Sweden	*Motor*	Jan. 10 1945
11	The Volvo PV444 Saloon	*Motor*	March 1 1950
13	Volvo 444 Saloon Road Test	*Autocar*	May 21 1954
18	The Volvo 444K Road Test	*Motor*	April 4 1956
22	Swedish Invasion - PV444 Road Test	*Sports Cars Illustrated*	Sept. 1956
28	Volvo PV444 California Saloon Road Test	*Autocar*	Sept. 7 1956
32	Sweden's Hot Volvo Road Test	*Motor Life*	Oct. 1956
34	Volvo PV444 Road Test	*Autocar*	June 28 1957
36	Volvo PV444L Road Test	*Road & Track*	Sept. 1957
38	Volvo PV444 Road Test	*Motor Trend*	Sept. 1957
40	A Hotter Volvo	*Sports Cars Illustrated*	Oct. 1957
42	Guide to Volvo Road Test	*Motor Lift*	Nov. 1957
44	How-to Tour in a Tyke	*Foreign Cars Illustrated*	April 1968
48	Rambler vs. Volvo Comparison Test	*Hot Road*	Sept. 1958
54	4 Speed Volvo Road Test	*Road & Track*	Oct. 1958
56	Volvo: Swedish, But no Iceberg Road Test	*Sports Car World*	April 1959
60	Servicing the Volvo	*Small Car Parade*	May 1959
64	Tractable Tiger	*Speed Age*	June 1959
67	Volvo 60hp Road Test	*Track & Traffic*	May 1961
68	Volvo PV544 "Sport" Saloon Raod Test	*Car South Africa*	June 1961
72	Volvo 544 Sports Road Test	*Sports Car Graphic*	April 1962
75	Volvo 544	*Track & Traffic*	Nov. 1963
79	Volvo PV544 Road Test	*Road & Track*	Nov. 1963
83	Volvo 544 Road Test	*Road Test*	May 1965
90	Volvo PV544 Owner's Report	*World Car Guide*	May 1970
95	Super Swede	*Classic and Sportscar*	Feb. 1983
96	Solid Sense - Volvo PV544 vs Saab 96 Comparison Test	*Popular Classics*	April 1991

ACKNOWLEDGEMENTS

At Brooklands Books, we are never happier than when motoring enthusiasts actually ask us to produce a title which is not yet in our lists. This is such a book and came about from a suggestion by Svenska Volvo PV Klubben, which in turn came to us by way of Thomas Salomonsson of A B Lafri the well known booksellers of Karlskrona Sweden. The club not only instigated the idea for the book but went even further by organising the photograph for our front cover. Johan Skarner took this lovely picture of Lars Hjertberg's 1952 PV444 DS. We are grateful, too, to motoring writer James Taylor for his brief introduction below.

The aim of this book is to gather together a selection of stories which will provide our reader with an overall picture of the life and times of the PV444 and PV544 models. In this aim, we have been greatly helped by the original publishers of the copyright material which the book reproduces. We are sure that Volvo enthusiasts will endorse our thanks to the managements of *Autocar, Car South Africa, Classic and Sportscar, Foreign Cars Illustrated, Hot Rod, Motor, Motor Life, Motor Trend, Road Test, Road & Track, Small Car Parade, Speed Age, Sports Car Graphic, Sports Cars Illustrated, Sports Car World, Track and Traffic,* and *World Car Guide*.

R.M. Clarke

INTRODUCTION

It was the PV444 and PV544 Volvos which really established the Swedish marque's reputation outside its home country: earlier models had been sold in Scandinavian countries, but with the PV444, Volvo broke into the major markets of Europe and the USA.

The PV444 was in fact announced in 1944, but a shortage of sheet steel meant that full production did not get under way until 1947. Such was the demand, however, that a thriving black market grew up around the trickle of earlier cars which came from the Gothenberg factory.

That demand should have surprised no-one. Volvo had deliberately made the PV444 smaller than any of their previous models, and had priced it to sell widely. And, even though the early 1940s American styling was anachronistic by the time the car was announced, the technical specification was remarkably advanced. Consider this: in 1944, the PV444 had integral construction, independent front suspension, coil springs and telescopic dampers all round, and a laminated windscreen as standard!

Over the next 14 years, the PV444 changed gradually, gaining more powerful engines, a larger glass area, a larger boot and a collection of other minor improvements. Yet its excellent roadholding and performance remained competitive throughout and, when Volvo introduced it to the US market in 1956, they were able with some justification to describe it as a sports sedan.

The 1958 PV544 Volvos which replaced the PV444s were even better. While retaining the basic shape of the older model, the car's manufacturers had given the PV544 more modern styling touches, a completely new interior, and the option of even more powerful engines. Over the next few years, the PV544 made a name for itself in motor sport and, kept abreast of its rivals by regular power increases, it remained a strong seller until it was taken out of production in 1965.

In 21 years of production, 440,000 PV444/544 models were made, both two-door saloons (there were never any four-doors) and Duett station wagons and vans. The cars remain enormously popular with enthusiasts - and deservedly so. I am delighted that there is, at last, a Brooklands Book about them.

James Taylor

JOSEPH LOWREY HAS

A DAY OUT IN A VOLVO

AROUND SURREY IN A SWEDISH SMALL CAR OF ADVANCED DESIGN

OWNER AND OWNED.—Dr. Lennart Hesselvik with the 1.4-litre Volvo which he is using for a visit to this country.

AMIDST the flood of British cars with which fugitives from austerity are packing the cross-Channel steamers, there are even now modest numbers of vehicles making the opposite journeys. Despite awful warnings, foreign visitors are coming to England again, and a recent caller at "The Motor" offices was the Swedish owner of probably the first post-war Volvo car to be brought into Britain.

Described in "The Motor" of January 10, 1945, the Swedish Volvo 1.4-litre car is an entirely new design which is only now beginning to reach private owners. Styled on American lines, but of popular European dimensions, it has an all-steel stress-carrying body supported on four coil springs, independent suspension being used at the front. For power unit, a four-cylinder engine of 75 mm. by 80 mm. is teamed up with an orthodox three-speed synchromesh gearbox.

Having transferred our visitor's car from the street, where it was collecting an alarmingly large crowd, to the comparative peace of the Temple Press car park, we indulged in a happy orgy of peering, probing and photographing. After which we were successful in persuading the car's owner that it would be a good thing to take the car for a run out of London on some convenient day, and an appointment was duly arranged.

Just where to take the car presented a pretty problem. With normal road-test cars the question of route hardly arises, being largely determined by the availability of performance-testing facilities. In this case, however, the car was still too new for full performance testing, our only demand in that direction being for a clear road on which to clear up the owner's doubts concerning fuel consumption. There was, on the other hand, an unusual consideration, that Dr. Lennart Hesselvik had a natural desire to see as much as possible of England.

The job of courier to foreign visitors has much to recommend it, but is a little outside the usual routine of a Technical Editor. However, it transpired that Surrey had hitherto escaped our friend's notice, so in the peace of a weekday morning the Volvo was headed towards the country south-west of London.

Lacking the boulevards of Paris, London, nevertheless, has a very pleasant and useful through road in the Embankments along many miles of the river's north bank. As introduction to a test car, it offers heavy traffic at the approaches to Blackfriars Bridge, brief open stretches, a roundabout and some really rough going succeeding one another as the road winds westwards. Traffic revealed that the Volvo did not love Pool petrol, the throttle requiring delicate treatment at speeds below 30 m.p.h. in top gear if pinking was to be avoided. The rougher pieces of road, however, were glossed over by the hydraulically damped coil springs, the car riding comfortably yet without any exaggerated softness.

London Features

Stockholm, with a population of nearly half a million, is no small town, but London feels very vast to a Scandinavian visitor. It has its interesting features, the big riverside power stations being unfamiliar to a visitor more used to hydro-electric generating plant. But by the time Putney Heath was reached, with the sun shining and swans on the lake, open country seemed overdue, and it was hard to admit that the built-up Kingston By-pass still remained to be traversed.

Leaving London along A3, Esher always seems to mark the beginning of real country. Running at around 50 m.p.h., the Volvo had a useful margin of power underfoot to maintain speed on hills, and little noise ever intruded upon its effortless progress. Left-hand drive, as always, proved a slight handicap on British roads, visibility being poor for overtaking if it became necessary to come close behind another car. By contrast with some left-hand-drive cars, however, the moulded-rubber gearbox cover provided very comfortable steadying for the right foot, simplifying delicate operation of the organ-pedal accelerator.

Driving visibility through the raked V-screen is good, although the bonnet is not unduly low. It may be personal preference, but I found I liked the rib down the centre of the bonnet better than the bare expanse of similarly shaped metal over which the drivers of some 1947 cars face the world. I do not demand "cherries on sticks" on the wings of a car, but I do not like sitting behind too featureless a front end.

The first real detour was made a few miles short of Guildford, when a signpost to Merrow reminded me of our proximity to Newlands Corner. On a Tuesday morning only one other car had stopped to admire the magnificent southward view. St. Martha's Chapel, on

AMERICAN STYLE.—This picture of the steering wheel and facia panel of the Volvo clearly shows the marked transatlantic influence on the design.

acceleration, the fuel consumption remains notably low.

One feature of Dr. Hesselvik's car, which suits Swedish winter conditions, was a roller blind in front of the radiator, controlled from the dashboard. It proved possible to blank the radiator almost completely without any boiling occurring, but on a hot day this fuel-saving method led to the engine showing the Pool petrol vice of running on after the ignition had been switched off. Starting, too, was less certain with a really hot engine, unless the throttle was opened fully.

Many would-be Swedish visitors to England have been put off by stories of cool welcomes and bad feeding. The Bush Hotel at Farnham was not entirely a casually chosen lunch place, although we made no advance booking, but it certainly incited one visitor to remark on what wrong impressions of England were prevalent abroad. Petrol allowances were, at that time, grossly inadequate for anyone wishing to see and spend money in Britain during a brief stay unless he was able to claim supplementary coupons as a business visitor.

Cricket on the Green

With a short afternoon available for showing off " a bit of England," we set off again from lunch to follow a zig-zag course and see what turned up. Cut Mill Ponds, a string of artificial lakes which stretch down a wooded valley from Hampton Park, was the first stop, trees providing welcome shade while fuel-consumption-test equipment was stripped off the car. Then, skirting Elstead, we headed for the typically English village green at Tilford, finding cricket in progress even on a weekday, though on this occasion the players were local schoolchildren.

Frensham Great Pond is once again a yachting centre after being drained during the war years, but to a Swedish visitor the half-mile expanse of water looked anything but great. On the way up towards Hindhead, however, I scored a greater success by turning right on to the unsigned lane which leads across Whitmoor Vale. From a scenic point of view, this particular road can be

the nearby hill across which the Pilgrims' Way once ran, was barely visible through the trees which surround it, and heat haze almost concealed the South Downs, but, even so, the outlook over the wooded country of Surrey and Sussex was truly magnificent. Incidentally, too, while the car was stopped Dr. Hesselvik was able to have his first look at the interior of an R.A.C. telephone box.

Guildford is a busy town and traffic congestion is not eased by tree-carrying trailers in narrow roads. Nevertheless, we gave the by-pass a miss, choosing rather to potter down the steep cobbled High Street, over which projects the fine old clock. Then up on to the Hog's Back road, which follows the crest of the chalk ridge towards Farnham, a stretch of road which is open and almost clear of houses, to make a few fuel-consumption checks.

Testing M.P.G.

Last year, after careful study of sundry proprietary fuel-consumption testing outfits and examination of a wide variety of cars, we planned our own equipment and had it built up by the North Downs Engineering Co It may look fearsome, but in its present form it really works, and so far no British, French or German car has defeated its universal fittings; the Volvo called for a little thought, but the engine is quite accessible when the forward-hinged bonnet is open, and a leak-proof installation was achieved with little removal of skin or paint.

The reason for making quite laborious fuel-consumption tests was, quite simply, to check up on some rather striking manufacturer's claims. The results of our tests are not submitted as comparable in precision with our regular road tests, there being modest possibilities of error in distance recording or due to an undulating road carrying normal traffic. Nevertheless, they are almost certainly within 2 m.p.g. of the figures which would be obtained under stricter conditions.

Any car which can cover 40 miles on a gallon of petrol fully deserves the adjective economical, and the Volvo must certainly be so described. Quite surprising is the ability to better 35 m.p.g. at a speed of 50 m.p.h., a feat for which an efficient engine and an aerodynamically clean body (somewhat festooned with pipes and wires during our tests) must share the credit. Even with use of the pump-type carburetter for frequent

DERIVATIVE.—These pictures show that the engine design is similar to that of another popular European car. The whole unit is mounted unusually far forward and lies between the suspension wishbones and coil springs.

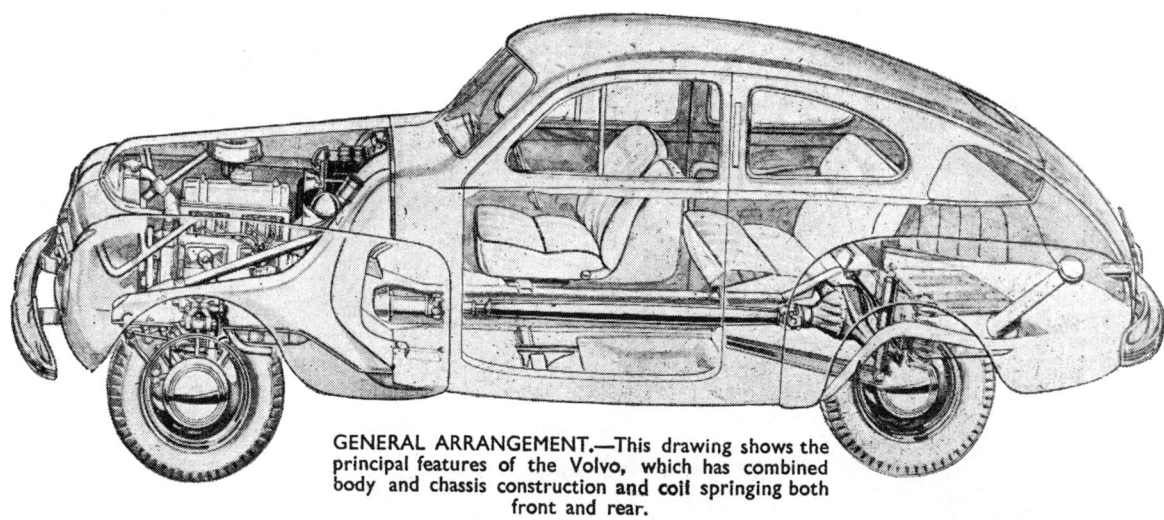

GENERAL ARRANGEMENT.—This drawing shows the principal features of the Volvo, which has combined body and chassis construction and coil springing both front and rear.

guaranteed to ring the bell at any time of year, and it also evoked two interesting comments on contrasts between British and Swedish roads.

Sweden the British think of as a land of wooded hills, yet this hilly and wooded road surprised Dr. Hesselvik. The steep hills on some British roads surprised him, for in Sweden gentler grading has to be used if traffic is not to be held up by the snows of winter. The complete canopies of foliage which turn many lanes into tunnels were another surprise, for although gravel roads can be very smooth, they would become hopelessly waterlogged at times if allowed to become so overgrown.

Hindhead is too well known to need much mention, the main London-to-Portsmouth road giving such magnificent views. We parked the Volvo for a few minutes to climb up past the stone "erected in detestation of a murder" to the crest of Gibbet Hill, then, after a quick glance at the famous all-round outlook, we descended and set course for London. It was still too early for tea when we passed the old Refectory Barn at Milford, so we pressed on, stopping only for fuel at the little garage in Peasmarsh, where Douglas Tubbs found his veteran Gobron Brillie.

Even a garagiste whose week-ends are devoted to Alfa Romeos and Allards found the Volvo surprising. The very name is unfamiliar in this country, as are the rather American exterior lines, while study of the well-equipped plastic facia is apt to lead to comments like: "Can you really play that Würlitzer?" Having overcome initial surprise, however, the interior of the car, with its light-green fabric trimming, cream plastic facia and polished-aluminium fittings appears bright and neat. The two-spoke steering wheel, with horn ring, is comfortable to use, the instruments are readily legible and the seats (with backs hinging diagonally to give access to the rear compartment) are comfortable for an all-day drive.

International Influences

Although made by an independent and old-established Swedish company, the Volvo shows some German influences in its power-unit design, while components of British and American manufacture are incorporated. The result is a unique-looking car, modern, viceless and attractive, but one which displays less individuality of character than might be expected from its background. A good car, filling the needs of many people and priced at under £400 in its homeland, it will undoubtedly be popular and should establish Sweden as a car-manufacturing country. But local conditions have not led to the evolution of a car which can teach the world's designers any new lessons.

Refuelled, we abandoned the A3 road by which we had left London, crossing the Wey to the villages of Shalford, Albury and Shere, following the valley road below the North Downs. Once again, though, we were tempted to turn aside and tackled the bottom-gear climb over the crest of the hills to Effingham before time relentlessly tied us to the main road through Epsom to London.

Reluctantly relinquishing the wheel of the Volvo, I returned to my hard-used British car, pondering how often different engineers quite independently reach similar solutions to their problems. Working in a country with little automobile-engineering background, a country where one needs a permit to buy a car even though petrol is unrationed, a country where gravel roads and cold winters are commonplaces, the Volvo engineers have produced a car almost precisely in line with the contemporary products of other countries.

VOLVO 444 SPECIFICATION

Engine Dimensions:	
Cylinders	4
Bore	75 mm.
Stroke	80 mm.
Cubic capacity	1,410 c.c.
Piston area	27.4 sq. ins.
Valves	Pushrod o.h.v.
Compression ratio	6.5

Engine Performance:	
Max. b.h.p.	43
at	4,000 r.p.m.
Max. b.m.e.p.	123 lb./sq. in.
at	2,000 r.p.m.
B.H.P. per sq. in. piston area	1.57
Peak piston speed ft. per min.	2,100

Engine Details:	
Carburetter	Carter downdraught
Ignition	Autolite coil
Plugs	Bosch
Fuel pump	Mechanical
Fuel capacity	8 gallons
Cooling system	Pump and fan
Electrical system	6-volt, C.V.C.

Transmission:	
Clutch	Single plate, 8 ins.
Gear ratios: Top	4.7
2nd	7.6
1st	16.8
Rev.	15.25
Propeller shaft	Spicer
Final drive	Spiral bevel

Chassis Details:	
Brakes	Lockheed hydraulic
Brake drum diameter	9 ins.
Suspension, front	I.F.S., Wishbones and coil springs
Suspension, rear	Torque arms and coil springs
Wheel type	Pressed steel
Tyre size	5.00 x 16
Steering wheel	Twin spoke

Dimensions:	
Wheelbase	8 ft. 6¼ ins.
Track, front and rear	4 ft. 3 ins.
Overall length	12 ft. 2 ins.
Overall width	5 ft. 1 in.
Dry weight	19 cwt.

Performance Data:	
Piston area, sq. ins. per ton	28.8 sq. ins.
Top gear m.p.h. per 1,000 r.p.m.	16.4 m.p.h.
Top gear m.p.h: at 2,500 ft. per min. piston speed	78 m.p.h.
Litres per ton mile, dry	2,700

Fuel Consumption:	
At steady 30 m.p.h.	
Eastwards	39.5 m.p.g.
Westwards	41.3 m.p.g.
MEAN	40.4 m.p.g.
At steady 40 m.p.h.	
Eastwards	39.1 m.p.g.
Westwards	41.3 m.p.g.
MEAN	40.2 m.p.g.
At steady 50 m.p.h.	
Eastwards	33.0 m.p.g.
Westwards	39.1 m.p.g
MEAN	36.0 m.p.g.

A Post-War Small Car from S

A full description of t
1.4-litre Volvo, which
an interesting blend
European and U.S.
design practice

NEAT WORK.—The front cross-member, engine mounting and I.F.S. system are built up in one unit as shown in this photograph. The transatlantic derivation can be clearly seen.

SCALED DOWN. might well be mi U.S.A. car. In ac scaled down in a proportion both track and height, base being rather at 8 ft. 6 ins. The g is a highly attractiv

FORWARD MOUNTING. — This picture shows how engine mounting in made on the front springing sub-component and the engine itself is brought well ahead of the wheel centres. The down-draught carburetter is provided with a very large air cleaner.

ALTHOUGH by no means the first post-war model to be described in the British technical Press, the 1.4-litre Volvo is the first post-war small car to be described which has in addition been designed throughout since 1939. For this reason it must command considerable interest, and it is desirable to expand the information given with the picture published in "The Motor" of October 18.

It should, perhaps, be mentioned that Sweden has a small but well-established motor industry, of which the two leading companies are Scania Vabis and Volvo. The latter has been making passenger cars for many years, and in 1938 it entered into an agreement with the Budd Corporation, of the U.S.A., to use the body-cum-chassis patents owned by that concern. Budd have numerous European connections,

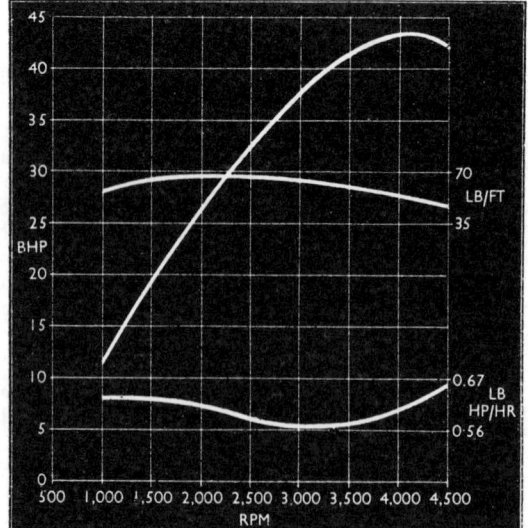

PERFORMANCE CHARACTERISTICS.—Despite a short stroke of 80 mm. designers of the Volvo have concentrated on obtaining good power and low engine speeds. The peak power is obtained at 4,000 r.p.m. as shown in the above curves, which also indicate the good full-throttle consumption of the unit.

and Citroën were amongst the first to take out a licence from them when they introduced their front-wheel-drive model in 1934. Whereas the Citroën and various other companies in Germany and the U.K. have put their own design interpretation upon the Budd system, the styling as well as the engineering of the Volvo are clearly derived from the far side of the Atlantic.

The pictures in this article are, in fact, a most interesting "tie-up" with those published in "The Motor" of January 3 and the article dealing with 1942 American cars. On the Volvo we see the characteristic wide, flat slot for air entry to the radiator, the inbuilt head lamps, the sharply sloping screen, and the easy taper on the stern of the car.

The close-up of the interior reinforces this alliance of Continental size with U.S.A style. The cloth seating and instrument panel are entirely out of line with, shall we say, British small-car layouts. The speedometer needle moves through the arc of a circle across a large, indirectly illuminated dial, and beneath are rectangular slots in which are partitioned the gauges for the other services. Below the dials is a grill for heater or radio, whilst on each side are large glove boxes with pull-down doors. The steering wheel, which has a plastic

eden

rim, also incorporates the horn ring, which, having originated in Europe, has now been widely adopted in U.S.A., so is thus making in effect a return circuit

In order to keep the overall height down, a deep tunnel in the centre of the car has had to be incorporated, but the floor is flat on each side of it. Stiffness is provided by the combined body and chassis construction previously mentioned. This is shown in the photograph of the front end, in which details not to be missed are the large diameter of the tubular cross-member and the deep section of the frame which runs forward from the dashboard.

Turning now from these general comments to some of the more technical details, it is interesting to observe that the independent front suspension linkage is built up on to a channel section cross-member, which acts as a further stiffener to the front end of the frame. Moreover, the rubber engine mountings are attached directly to the suspension component, as can be seen clearly in one of the illustrations. The springing unit is again highly reminiscent of the U.S.A. and, in particular, General Motors practice, there being two wishbones of unequal length and coil springs.

The rear-springing layout is almost identical to that employed on the Oldsmobile, although on a smaller scale. That is to say, there is a conventional rear axle with a spiral bevel reduction gear, the springs being coils. This axle is located transversely by a Panhard rod, and fore and aft by two long arms, which are inclined towards the centre line of the frame. These also take the driving torque through the medium of rubber bushes, probably of the Harris-flex type. The transmission itself is quite conventional through Spicer needle-bearing-type universals. three-speed gearbox and a single-plate clutch. The power unit, on the other hand, shows many features of distinct interest, particularly to British designers' now freed from the bias of piston area taxation.

Although only 1,410 c.c. capacity, the bore is 75 mm., the stroke only 80 mm., giving a s/b ratio of only 1 : 1066. Despite this short stroke, maximum power (43 b.h.p.) is developed at only 4,000 r.p.m., and the output per square inch of piston area is held down to the relatively modest figure of 1.57 b.h.p. Nevertheless, the short stroke does show its particular advantage by providing the very low maximum piston speed of only 2,100 ft. per minute at the peak of the horse-power curve. This aspect of the engine design is in harmony with the high gear of the car, and whereas comparable English models are geared to give 3,100 r.p.m. and 2,000 f.p.m. at 50 miles per hour, in the case of the Volvo the relevant figures are 2,700 r.p.m. and 1,420 f.p.m. respectively

Despite these facts, the road performance should be quite good, as the weight has been held down to the commendable low figure of 17½ cwt. The dimensions have not been squeezed, for the wheelbase of 8 ft. 6½ ins and a track of 4 ft. 3 ins. are both within an inch of British averages, although the car as a whole

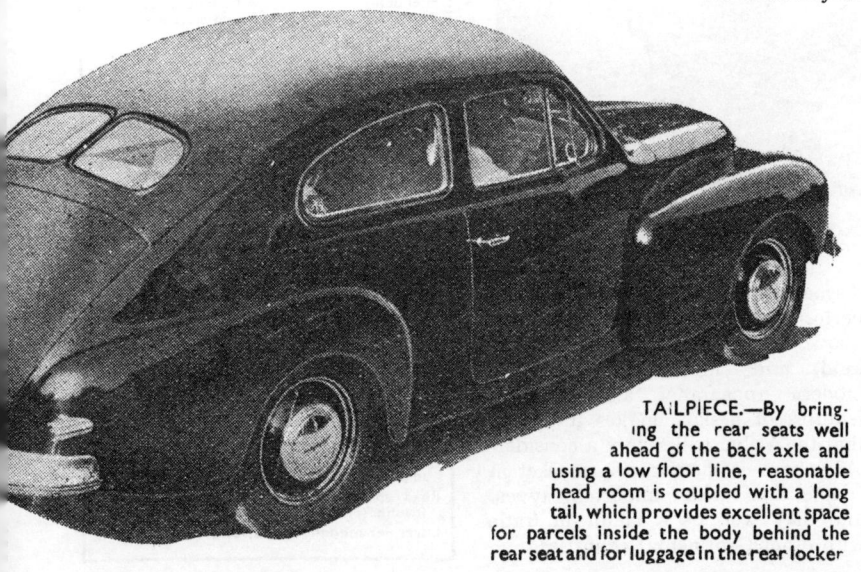

TAILPIECE.—By bringing the rear seats well ahead of the back axle and using a low floor line, reasonable head room is coupled with a long tail, which provides excellent space for parcels inside the body behind the rear seat and for luggage in the rear locker

9

A Post-War Small Car from Sweden—Contd.

weighs 6 cwt. less than the British Twelves road-tested by "The Motor" in 1939. In consequence, the litres-per-ton-mile figure comes out at 2,930, and even with two up and fuel, an acceleration of 10-30 m.p.h. in about 10.5 secs. should be obtained on top gear. This excellent combination of low piston speed and relatively good acceleration derives directly from a proper proportion of weight-to-piston area, the Volvo figure being 71.5 lb. per square inch, as compared to an average of 100 lb., which is the normal figure for cars of this size and type.

A 70 m.p.h. Maximum

No figures are available giving the performance of the car in the upper speed ranges, but it is significant that the peak of the power curve corresponds with a road speed of 66 m.p.h., so that a flat-out maximum of 70 m.p.h. is a reasonable assumption. In this respect the designer appears to have deliberately chosen to sacrifice maximum in order to achieve good low end performance.

ROUGH, AND FINISHED.—Right is seen the welded pressed steel assembly which fulfils the dual purpose of body and chassis and is made under Budd patents. Below is the interior view of the car which shows that in detail as well as in general form U.S.A. influence has been predominant.

In view of the fact that the horse-power and frontal area figures are very close to those of the Lancia-Aprilia, and that the drag of the Volvo should be little, if any, greater than that of the Italian model, a higher rear-axle ratio should permit an increase in the maximum to a round 80 m.p.h., as even at this speed the piston velocity would be less t' in 2,600 ft. per minute. Even discounting such speculations, the performance may be considered above the average of cars of this kind, and, in combination with modern appearance and practical features of design, makes it appear that the Volvo will secure a considerable share of the Swedish market in competition with imported types, and may well be a factor in trade with other Continental countries.

VOLVO DATA PANEL

	TYPE 444
Present tax	£17 10s.
Cubic capacity	1,410 c.c.
Cylinders	4
Valve position	Overhead
Bore	75 m.m.
Stroke	80 m.m.
Comp. ratio	6.5 to 1
Max. power	43 b.h.p.
at	4,000 r.p.m.
Max. torque	70 lb./ft.
at	2,000 r.p.m.
H.P. : Sq. in. piston area	1.57
Weight : Sq. in. piston area	71.5 lb.
Piston speed (ft./min.) at max. h.p.	2,100
Carburetter	Downdraught
Ignition	Bosch
Plugs : Make and type	Bosch
Fuel pump	A.C. Mechanical
Clutch	Single-plate 8 ins.
First gear	16.8
Second gear	7.6
Top gear	4.7
Reverse	15.25
Propel'er shaft	Needle bearing Spicer
Final drive	Spiral bevel
Brakes	Lockheed
Drums	9 ins.
Suspension	Independent front, coil springs, torque arms rear, coil springs
Wheelbase	8 ft. 6¼ ins.
Track, front	4 ft. 3 ins.
Track, rear	4 ft. 3 ins.
Overall length	12 ft. 2 ins.
Overall width	5 ft. 1 in.
Weight—dry	17.5 cwt.
Tyre size	16 x 5.00
Wheel type	Disc
Fuel capacity	7.5 gallons
Electrical system	Bosch 6-volt

TOP GEAR FACTS:

Engine speed per 10 m.p.h.	610 r.p.m.
Piston speed per 10 m.p.h.	320 ft./min.
Road speed at 2,500 ft./min. (piston)	78 m.p.h.
Litres per ton-mile	2,930

The VOLVO PV 444 SALOON

A Straightforward Modern Four-seater Car Combining Economy with Brisk Performance

A UNIQUE position among European cars is occupied by the Volvo PV 444, a model of Swedish design and construction which we were recently able to sample for several days in its home country. It has been designed especially to suit a home market whose buyers, hitherto accustomed to roomy and generously powered automobiles of American type, are now obliged by increased purchase and running costs to use smaller, more economical vehicles.

In dimensions the Volvo is very typically European, with its four-cylinder engine of 1.4 litres displacement and its bodywork proportioned to accommodate four people and their luggage in comfort: its most unorthodox dimensions are perhaps the rather unusually generous wheelbase and ground clearance, dimensions well suited

SMOOTH ENTRY.— Aluminium replaces chromium plate for much ornamental metalwork, and the car's clean lines allow 74 m.p.h. to be obtained on only 40 b.h.p. A single steel pressing forms the sweeping roof.

to a country of slippery winter roads and of vast, thinly populated districts.

In appearance, the car shows signs of strong transatlantic influence on both interior and exterior styling; although in fact unlike older designs carrying the same maker's name the PV 444 is not the work of designers from outside Sweden. Pressed steel has been used to produce an integral body of sensibly simple shape, and the internal furnishing uses plastics and fabric in the American manner.

Comfort

On the road, the Volvo continues to present a mixture of American and European characteristics, but it is a fortunate mixture partaking of many of the best qualities from old and new worlds. Nonessential items of equipment calculated to add unnecessarily to weight or wind resistance have been consistently avoided, and it has in consequence proved possible to offer lively performance yet to keep fuel consumption down to a very economical level.

The steel body, which also serves as chassis frame, makes no attempt to accommodate more than four passengers but offers very excellent comfort for that number. Head room is unusually generous over the front seats, and more than adequate in the back of the car despite sweeping tail lines. Elbow room is equally ample, and the rear passengers have almost unrestricted foot space below the front seats.

The seats are set at such a height above the floor that both driver and passengers are pleasantly upright, the former enjoying excellent forward vision over the short bonnet. The only criticism which may reasonably be levelled against the body is that entry and exit are slightly less easy than is sometimes the case, a small sacrifice in this respect having been perhaps inevitable to secure roominess and low weight. The floor is flat on either side of the considerable propeller shaft tunnel, but it is recessed appreciably below the level of the door sills, the front-hinged doors leave only limited space between their pillars and the front seat cushions through which feet may be passed, and even with diagonal hinges on the seat backrests access to the rear seats of a two-door body cannot be ideally easy. Once entered, however, the seats are exceptionally comfortable, screwed adjusters for backrest angle supplementing the usual sliding adjustments provided for the individual front seats.

Accommodation for luggage in the boot is very generous, both beside the spare wheel and on a shelf above it, and there are sufficient places for stowing minor packages inside the car. Lockers are provided on each side of the facia panel, there is a large shelf below the sloping rear window, and parcels put under the front seats do not intrude upon footroom for the rear passengers.

Supported on coil springs at both front and rear, the car rides very comfortably although not with quite the softness of certain present-day designs. The minor corrugations of earth-surfaced roads or of stone setts are completely absorbed, and the amount of body movement induced by larger bumps is quite moderate. It may be detected by eye that the ride is not strictly flat over some surfaces, but nevertheless the rear seat passengers enjoy almost as comfortable a ride as do those in front: doubtless low friction and unsprung weight, resulting from the elimination of leaf springs, contribute to the lack of shock experienced by back-seat occupants.

Comfortably sprung, very generously provided with ground clearance and internal headroom, and with a track of only 51 inches at the wider (rear) axle, the car might perhaps be expected to roll easily on corners: in fact, it is realized in the negotiation of winding roads that this

DOUBLE SAFETY.—Hinged at the front to eliminate risk of its opening while the car is running, the bonnet is automatically held by a positive catch when raised.

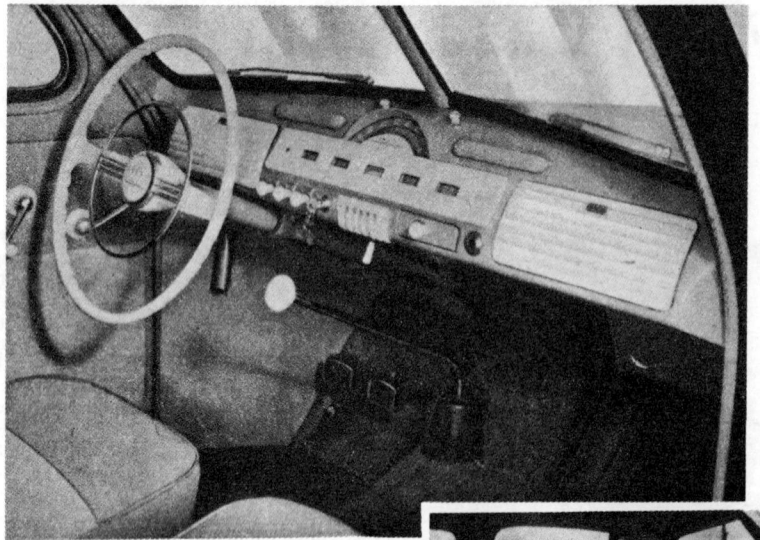

STYLED IN PLASTIC.—The painted metal facia panel has plastic instrument surrounds, and carries ammeter, oil pressure gauge, odometer, thermometer and fuel contents gauge below the central speedometer. (*Inset*) Diagonally folding backrests give access to the rear seats.

Volvo Road Test—Contd.

is scarcely so at all, the car remaining at all times on an almost level keel.

The rear springing arrangements provide a measure of stabilisation against roll, the torque arms which share with a Panhard rod the duty of axle location being swept inwards but only to points spaced on either side of the propeller shaft: roll stiffness here is moderated to the required level by the insertion of thick rubber pads at the points where U-bolts clamp the axle to the top-hat section torque arms. A matching degree of roll stiffness is imparted to the independent front wheel springing system by an orthodox rubber-bushed torsion-bar stabiliser.

Control

Lightness of handling has obviously been sought, and has been successfully attained with the very normal steering ratio of 3¼ wheel turns between the extremes of a satisfactorily compact lock. At no time is much effort needed on the steering, which at normal speeds is finger light and shows the minimum of road reaction yet which allows the car to be placed on the road with satisfactory precision. No castor angle or trail is used in the steering, which is little affected by tramlines or other ruts: normally there is a small but adequate amount of self-centring action, but for fast cornering something a little more vigorous might be advantageous.

The brakes do not feel quite so feather-light to the touch as does the steering, but nevertheless respond progressively and powerfully to the driver's requirements. They check the car in a straight line on slippery surfaces, and being identical to those fitted on a popular American automobile of double the Volvo's engine displacement should be durable and free from fade.

Light to operate, but perhaps a trifle quick in engagement, the clutch needs gentle treatment if a really smooth start from rest is to be made. The gearbox, controlled by an orthodox central lever, incorporates excellent baulk-ring synchromesh on the upper two of its three ratios: second gear, which is relatively high for a 1.4 litre car, falls short of the best modern standards of silence, but the helical-geared first ratio is commendably quiet for a non-constant-mesh gear.

Engine flexibility does not extend down to quite such a low road speed on the PV 444 as on some contemporary four-cylinder models, perhaps because of the high gear ratios used or in consequence of a generous carburetter choke size being employed. On a steep hill, the engine can be allowed to continue pulling in direct drive until the speed has fallen as low as 12 m.p.h., but appreciable hesitancy marred the top gear acceleration at 10 m.p.h. Above these rather low speeds, however, the power unit is extremely smooth.

Extremes of cold were not experienced during our testing of this car, although they are expected in its home country. After standing in the street through nights of moderate frost, however, the engine started at one touch of the starter knob, and with half choke the car instantly could be driven away normally.

At the brisk main road speeds which are natural to it, the Volvo is troubled by wind noise if the windows or hinged ventilation panels are opened slightly, but the car becomes incomparably quieter when these are closed. Body panels are damped internally by sheets of sound deadening material, and drumming is not prominent. The body is free from squeaks or rattles, but the doors only close securely when slammed very hard.

Performance

Over and above all questions of how the car behaves, however, there predominates the matter of what it does. Intended as the "people's car" of a reasonably prosperous European country, and exported only on a scale sufficient to pay for a few American and British fittings incorporated in it, the PV 444 is not a luxury model but is essentially meant to do a serious job of work.

That it is more than able to do any normal job is clearly evident from the figures published on the data page. It is a car which should give most owners 30 miles per gallon of fuel used, often more and seldom less unless driving conditions work the accelerator pump especially hard. Yet, with this order of economy, it will attain a genuine 74 m.p.h. on the level, and has acceleration well above the average for 1½-litre saloons on a top gear ratio so high that even maximum speed is within accepted continuous cruising limits of piston speed.

Avowedly designed as a small car with American characteristics the Volvo is an extremely creditable example of how excellently orthodox modern engineering can combine the conflicting qualities of speed and fuel economy, comfort and modest manufacturing cost. Already more than 8,000 black-painted examples of the type are in service in Sweden and elsewhere, and with the range shortly to be extended to include cars painted in colours and with "de luxe" body furnishing (at slight extra cost) the Volvo PV 444 should continue to reflect great credit upon the Swedish engineering industry.

UNORTHODOX CAPACIOUSNESS.—With the spare wheel held centrally under a shelf, the luggage locker is deep enough to take several large suitcases.

The Motor Continental Road Test No. 1C/50

Make: Volvo
Makers: Aktiebolaget Volvo, Göteborg, Sweden
Type: PV444

Dimensions and Seating

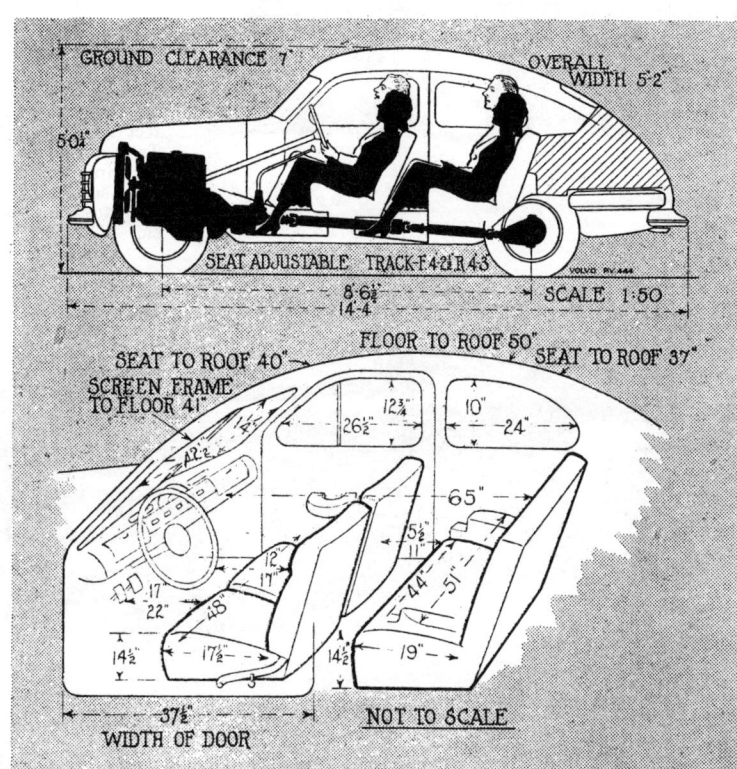

In Brief

Price Swedish Kr. 6490 = £448.
Capacity .. .. .. 1,414 c.c.
Unladen kerb weight .. .. 19 cwt.
Fuel consumption .. .. 33 m.p.g.
Maximum speed .. 74.1 m.p.h.
Maximum speed on 1 in 20 gradient .. .. 59 m.p.h.
Maximum top gear gradient 1 in 12.7
Acceleration,
  10-30 m.p.h. in top . 12.4 secs.
  0-50 m.p.h. through gears 17.6 secs.
Gearing 16.3 m.p.h. in top at 1,000 r.p.m. 78 m.p.h. at 2,500 ft. per min. piston speed.

Specification

Engine
Cylinders 4
Bore .. .. .. 75 mm.
Stroke .. .. .. 80 mm.
Cubic capacity .. .. 1,414 c.c.
Piston area .. .. 27.4 sq. ins.
Valves .. .. Pushrod o.h.v.
Compression ratio 6.4/1
Max power .. .. 40 b.h.p.
  at .. .. 3,800 r.p.m.
Piston speed at max. b.h.p. 2,000 ft. per min.
Carburetter .. .. Carter WO-618-S d/d
Ignition .. .. .. Coil
Sparking plugs .. 10 mm. Champion Y6
Fuel pump .. .. Mechanical
Oil filter .. Optional Fram by-pass
Electrical equipment .. Auto-lite, 6-volt

Transmission
Clutch .. .. Borg and Beck s.d.p.
Top gear (s/m) 4.55
2nd gear (s/m) 7.40
1st gear 14.7
Propeller shaft .. .. Hardy Spicer
Final drive .. (E.N.V.) Hypoid bevel, 9/41

Chassis
Brakes .. .. Wagner hydraulic
Brake drum diameter 9 ins. × 2 ins. width
Friction lining area .. .. 145 sq. ins.
Suspension :
  Front .. Coil and wishbone I.F.S.
  Rear .. Coil springs and rigid axle
Shock absorbers .. .. Delco telescopic
Tyres .. .. .. 5.00 × 16, 4-ply

Steering
Steering gear .. .. Ross
Turning circle .. .. 34 ft.
Turns of steering wheel, lock to lock .. 3¼

Performance factors (at laden weight as tested)
Piston area, sq. in. per ton 24.1
Brake lining area, sq. in. per ton .. 127
Specific displacement, litres per ton-mile 2,300
Fully described in "The Motor," January 10, 1945.

Test Conditions

Frosty, dry weather, with little wind. Smooth concrete surface. Swedish pump fuel (approx. 72 octane).

Test Data

ACCELERATION TIMES on Two Upper Ratios

	Top	2nd
10-30 m.p.h.	12.4 secs.	6.5 secs.
20-40 m.p.h.	10.7 secs.	6.6 secs.
30-50 m.p.h.	12.7 secs.	10.3 secs.
40-60 m.p.h.	15.0 secs.	—

ACCELERATION TIMES Through Gears

0-30 m.p.h. .. .. 6.8 secs.
0-40 m.p.h. .. .. 10.6 secs.
0-50 m.p.h. .. .. 17.6 secs.
0-60 m.p.h. .. .. 24.9 secs.
Standing quarter-mile .. 23.1 secs.

MAXIMUM SPEEDS
Flying Quarter-mile
Mean of four opposite runs .. 74.1 m.p.h.
Best time equals .. 74.4 m.p.h.

Speed in Gears
Max. speed in 2nd gear .. 52 m.p.h.
Max. speed in 1st gear .. 26 m.p.h.

FUEL CONSUMPTION
44.0 m.p.g. at constant 20 m.p.h.
41.0 m.p.g. at constant 30 m.p.h.
39.0 m.p.g. at constant 40 m.p.h.
35.5 m.p.g. at constant 50 m.p.h.
30.5 m.p.g. at constant 60 m.p.h.
24.0 m.p.g. at constant 70 m.p.h.
Overall consumption for 145 miles, 4.4 gallons, equals 33 m.p.g.

WEIGHT
Unladen kerb weight .. .. 19 cwt.
Front/rear weight distribution 53/47
Weight laden as tested .. 22¾ cwt.

INSTRUMENTS (calibrated in kilometres)
Speedometer at 30 m.p.h. .. 5% fast
Speedometer at 60 m.p.h. .. 5% fast
Distance recorder .. .. 1% fast

HILL CLIMBING (at steady speeds)
Max. top-gear speed on 1 in 20 .. 59 m.p.h.
Max. top-gear speed on 1 in 15 .. 53 m.p.h.
Max. gradient on top gear .. 1 in 12.7 (Tapley 175 lb./ton)
Max. gradient on 2nd gear .. 1 in 7.4 (Tapley 300 lb./ton)

BRAKES AT 30 m.p.h.
0.97 g. retardation (=31 ft. stopping distance) with 90 lb. pedal pressure.
0.60 g. retardation (=50 ft. stopping distance) with 50 lb. pedal pressure.
0.32 g. retardation (=94 ft. stopping distance) with 25 lb. pedal pressure.

Maintenance

Fuel tank: 7.7 gallons. **Sump:** 6½ pints, S.A.E. 20 summer, S.A.E. 10 winter. **Gearbox:** 1 pint gear oil, S.A.E. 90 summer, S.A.E. 80 winter. **Rear axle:** 2 pints E.P. gear oil, S.A.E. 90 summer, S.A.E. 80 winter. **Steering gear:** Gear oil. **Radiator:** 17 pints. **Chassis lubrication:** By grease gun every 1,250 miles to 23 points. **Ignition timing:** Static, T.D.C. **Spark plug gap:** 0.030 in. **Contact-breaker gap:** 0.018 in. **Carburetter main jet:** 0.070 in. **Valve timing:** I.O., 5° B.T.D.C.; I.C., 47° A.B.D.C.; E.O., 47° B.B.D.C.; E.C., 5° A.T.D.C. **Tappet clearances (hot):** Inlet, 0.008 in.; exhaust, 0.008 in. **Front wheel toe-in:** 5/64th in. at centre of tread. **Camber angle:** 0±¼°. **Castor angle:** 0° to −½°. **King-pin inclination:** 4½°. **Tyre pressures:** Front, 24 lb.; rear, 27 lb. **Brake fluid:** Delco 11. **Battery:** 6-volt, 85-amp./hour. **Lamp bulbs:** Head lamps, 35/35-watt Duplo; parking lamps, 1.5-watt; number plate lamp, 3-watt; stop/tail lamp, 20/3-watt; panel light, 2.4-watt; direction indicator, 3-watt; roof light, 3-watt; facia warning lights, 2.4-watt.

Ref. S/15/50

Unlike the main contours of the Volvo, the rear end of the front wings finishes in an abrupt vertical line. Protective rubbers are attached to the leading edges of the rear wings to prevent damage to paintwork. A jacking bracket can be seen below the door, midway along the wheelbase.

The rear bumpers sweep round to protect the wing panels. Twin tail lamps are built into the wings and a small reflector is placed on each side of the luggage locker lid.

The Autocar ROAD TESTS

No. 1532: VOLVO 444 SALOON

WITH the art of vehicle design developed to its present state it is not unusual for the qualities of a car to be assessed as either American or Continental, and in either case a person who is reasonably well informed about cars in general would understand what were the basic features of the car under discussion. Vehicles not designed in the United States or in the major car producing countries of Europe, such as Great Britain, France, Italy and Germany, may have characteristics that are perhaps less sharply defined as a basic type, but on the other hand embody desirable features that are found in vehicles designed on both sides of the Atlantic. A car in this category has recently been tested on the Continent by this journal; the Volvo 444 produced in Sweden by Aktiebolaget Volvo, who have been manufacturing cars since the middle 'twenties. They also make a larger car known as the Volvo Disponent, a range of commercial vehicles, and a van chassis with some components basically similar to those used in the 444. Both a standard and a de luxe saloon version of the 444 are available, the two vehicles differing only in minor details and fittings. It is the standard model that is the subject of this test by arrangement with the main Volvo distributors in Holland, N. V. Nederlandsche Bedrijfsauto-Import Mij., The Hague.

Briefly, the 444 combines the general comfort and convenience associated with a vehicle of transatlantic design with the handling qualities that are expected of a thoroughbred Continental car. Added to these qualities is a third, that of good general and detail finish and sound engineering, for which this car's country of origin has a reputation.

A picture of the overall efficiency of the complete vehicle can be visualized when it is realized that the engine, of under 1½-litre capacity, is capable of propelling the car—a comfortable four-five-seater weighing almost 22½ cwt in road test condition—at a mean speed of 76 m.p.h.: it also has an overall fuel consumption of 28 m.p.g. Further, the car is geared so that it will attain its maximum speed without the use of miles of unobstructed motor road. At the same time if long stretches of such road are available, the Volvo can cruise indefinitely at near maximum without showing signs of distress.

With a relatively low compression ratio of 6.5 to 1 the four-cylinder engine is smooth, starts easily and attains its working temperature very quickly, as it is fitted with a thermostatically controlled induction manifold hot-spot with an adjustment for summer and winter conditions, and also

From the front the Volvo is distinctive. The bonnet is pivoted at the front so that it opens up over the fixed portion formed by the front wings and grille. Over-riders are fitted to the bumper. All the fixed windows are mounted in rubber, although there is a bright strip covering the V joint in the centre of the windscreen.

Both front and rear seats are upholstered in striped cloth, while the whole of the floor is covered with rubber. Combined door pulls and arm rests are fitted to the doors, and side arm rests are provided for the rear passengers. There is an ash tray in the centre of the facia and for the rear passengers ash trays are provided at the front of the arm rests. The front seat back rests fold forward to aid getting into and out of the rear compartment.

a chain-operated roller blind in front of the radiator. If this last item of equipment is used—and intelligently used, it is of decided value towards engine efficiency and economy—it is necessary to watch the temperature gauge in order to guard against the engine overheating; a warning light to supplement the water temperature gauge would be a useful addition.

The transmission is orthodox with a dry single-plate clutch and three-speed gear box. The clutch is smooth in operation, has a satisfactory pedal travel, and is pleasantly light to operate. It is also well able to cope with rapid acceleration without excessive slip. Of the three well-chosen ratios in the gear box synchromesh is provided for top and second gears; these are controlled by a central lever on orthodox lines, which has a very clean action. The synchromesh is also positive and not easily beaten. In traffic and when starting from rest it is frequently necessary to use first gear, and it would be useful if synchromesh were provided on this ratio as well. The transmission generally is quiet, although a certain amount of gear box rattle could be heard if the car was held at maximum speed; on the car tested this could be eliminated if the driver rested his hand on the gear lever. A two-piece propeller-shaft transmits the drive to the hypoid rear axle, which is of British manufacture.

Suspension and Roadholding

The arrangement of well-damped coil springs at both front and rear results in a very comfortable ride for all occupants in a wide variety of road conditions ranging from high-speed motor roads to very rough stone setts. The roadholding qualities of the 444 are also of a very high order, a particularly reassuring feature in a country such as Holland, where the Road Test was performed, as there is often a very short distance between road and canal! The car tested was fitted with tyres of Dutch origin with an unusual ribbed tread with knife cuts around the periphery of the ribs, so that they "bare their teeth," as it were, under conditions of acceleration and braking. To complement the good roadholding qualities the car has another essential quality necessary to provide good directional stability, a satisfactory degree of understeer, a feature which, because of the weight distribution and general layout, does not noticeably change with variations in passenger loading.

The steering is light, well balanced and provides a useful sense of road feel without transmitting road shocks. It also has a satisfactory self-centring action, and, although there are $3\frac{1}{4}$ turns from lock to lock, the driver is not conscious of the need for excessive wheel movement in normal driving, while the steering is pleasantly light for manoeuvring in confined spaces.

Braking Behaviour

Hydraulically operated brakes have leading and trailing shoes at both front and rear. Under test conditions these recorded a satisfactory efficiency for quite moderate maximum pedal pressure; they also operate efficiently for normal check braking with a small applied load. No noticeable fade was experienced under the severe and specialized performance testing conditions, and the brakes remained perfectly balanced throughout the test distance of many hundreds of miles, although there was a very slight increase in free pedal travel. The brake pedal has a very satisfactory solid feel, and the hand brake, coupled to the rear wheels, is also effective.

General noise level in the 444 is commendably low. As regards mechanical noises, apart from the transmission noise already mentioned, there was a slight rear axle noise on the particular model tested, but there is very little wind noise and no body boom. The car is also well insulated from noises set up by road surfaces and is free from vibrations. There is very little tyre noise even when driving over rough surfaces, a feature no doubt influenced by the tread pattern, which presents a smooth ribbed tread under normal constant speed conditions. No tyre squeal was noticed in normal cornering.

Driver convenience is a marked feature of the interior layout; the driving seat is amply proportioned, and has a 19in cushion which gives plenty of support for the legs. The seat cushion itself is a little on the hard side, but not unduly so, and did not cause discomfort even after a long spell at the wheel. The back rest is also well arranged and is adjustable for rake by altering shims placed under the abutment screws. The relative positions of the steering wheel and pedals results in a very comfortable driving position, and very little adjustment of the driving seat is necessary for comfortable accommodation of drivers of widely differing proportions. An organ-pedal type of throttle control is used, and this is correctly positioned in relation to the brake pedals; in the left-hand-drive car tested the central tunnel provides a useful steady for the driver's right foot. The foot-operated dip switch to the left of the clutch pedal is also very nicely positioned; it is extremely compact, so that it does not obstruct an unnecessarily large area, or project a long way out from the rubber-covered floor.

From the driving seat the forward field of vision is satisfactory, although it is not possible to see the opposite side front wing. The windscreen pillars are fairly thick in the fore and aft dimension but are placed so that they do not noticeably obstruct the view. On the other hand, the arrangement of the small two-piece rear window, together with the position of the driving mirror, tends to restrict the rearward vision.

All the instruments are neatly grouped on the facia in front of the driver, where they can be clearly seen through the T-spoked steering wheel. They consist of a centrally placed speedometer with an oil pressure gauge and ammeter on the left and water temperature and fuel gauges on the right. The instrument lighting is good and a rheostat

VOLVO 444 SALOON

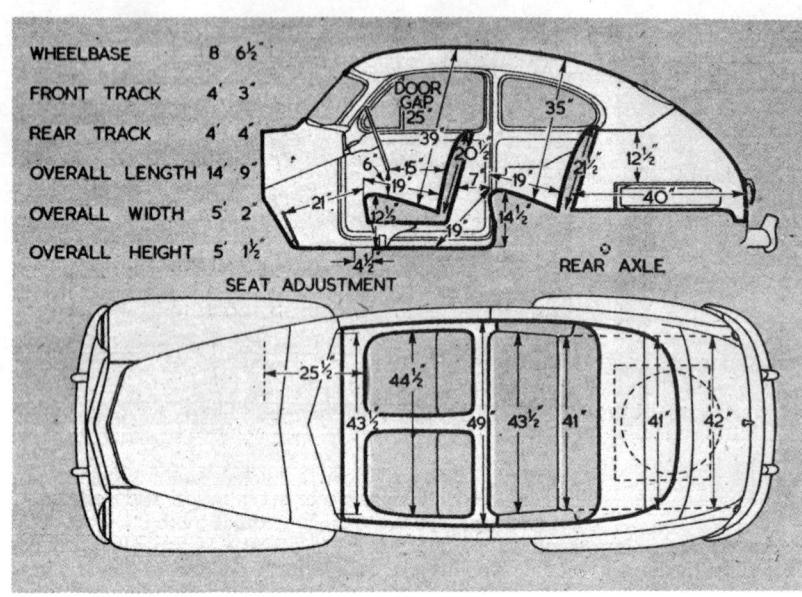

Measurements in these ⅛in to 1ft scale body diagrams are taken with the driving seat in the central position of fore and aft adjustment and with the seat cushions uncompressed.

PERFORMANCE

ACCELERATION: from constant speeds.
Speed Range, Gear Ratios and Time in sec.

M.P.H.	4.55 to 1	7.4 to 1	14.7 to 1
10—30	14.2	8.1	—
20—40	13.7	8.3	—
30—50	13.7	11.0	—
40—60	17.0	—	—
50—70	26.2	—	—

From rest through gears to:

M.P.H.	sec.
30	6.9
50	17.9
60	29.6
70	55.0

Standing quarter mile, 23.9 sec.

SPEEDS ON GEARS:

Gear	M.P.H. (normal and max.)	K.P.H. (normal and max.)
Top (mean)	76	122.3
(best)	78	125.5
2nd	50—58	80—93
1st	20—26	32—42

TRACTIVE RESISTANCE: 19 lb per ton at 10 M.P.H.

SPEEDOMETER CORRECTION: M.P.H.

Car speedometer	10	20	30	40	50	60	70	81
True speed	7	17	27.5	38	48.5	58.5	68.5	78

TRACTIVE EFFORT:

	Pull (lb per ton)	Equivalent Gradient
Top	170	1 in 13.1
Second	273	1 in 8.2

BRAKES

Efficiency	Pedal Pressure (lb)
82 per cent	80
73 per cent	60
48 per cent	40

FUEL CONSUMPTION:
28 m.p.g. overall for 300 miles (10.1 litres per 100 km.)
Approximate normal range 27-34 m.p.g. (10.5-8.3 litres per 100 km.)
Fuel, Dutch ordinary grade (approximately 79 octane).

WEATHER: Fine, dry surface; slight wind.
Air temperature 59 deg F.
Acceleration figures are the means of several runs in opposite directions.
Tractive effort and resistance obtained by Tapley meter.
Model described in *The Autocar* of March 5, 1954.

DATA

PRICE (in Holland), with two-door saloon body, 7,965 guilders = £748 at 10.65 guilders = £1. Not available in Great Britain.

ENGINE: Capacity: 1,414 c.c. (86.62 cu in).
Number of cylinders: 4.
Bore and stroke: 75 × 80 mm (2.953 × 3.15in).
Valve gear: Overhead; push rods.
Compression ratio: 6.5 to 1.
B.H.P.: 44 at 4,000 r.p.m. (B.H.P. per ton laden 39.3).
Torque: 67 lb ft at 2,200 r.p.m.
M.P.H. per 1,000 r.p.m. on top gear, 16.

WEIGHT (with 5 gals fuel): 19 cwt (2,130 lb).
Weight distribution (per cent): F, 51.5; R, 48.5.
Laden as tested: 22.4 cwt (2,510 lb).
Lb per c.c. (laden): 1.78.

BRAKES: Type: F, Leading and trailing; R, Leading and trailing.
Method of operation: F, Hydraulic; R, Hydraulic.
Drum dimensions: F, 9in diameter; 2in wide. R, 9in diameter; 2in wide.
Lining area: F, 65 sq in. R, 65 sq in (116 sq in per ton laden).

TYRES: 5.90—15in.
Pressures (lb per sq in): F, 21; R, 24 (normal).

TANK CAPACITY: 7¼ Imperial gallons.
Oil sump: 6½ pints.
Cooling system: 14 pints (plus 1¼ pints if heater is fitted).

TURNING CIRCLE: 33ft 6in (L and R).
Steering wheel turns (lock to lock): 3¼.

DIMENSIONS: Wheelbase: 8ft 6½in.
Track: F, 4ft 3in; R, 4ft 4in.
Length (overall): 14ft 9in.
Height: 5ft 1½in.
Width: 5ft 2in.
Ground clearance: 8in.
Frontal area: 20 sq ft (approximately).

ELECTRICAL SYSTEM: 6-volt; 85 ampere-hour battery.
Head lights: Double dip; 45-40 watt bulbs.

SUSPENSION: Front, Independent; coil springs and wishbones. Anti-roll bar. Rear, Coil springs and links.

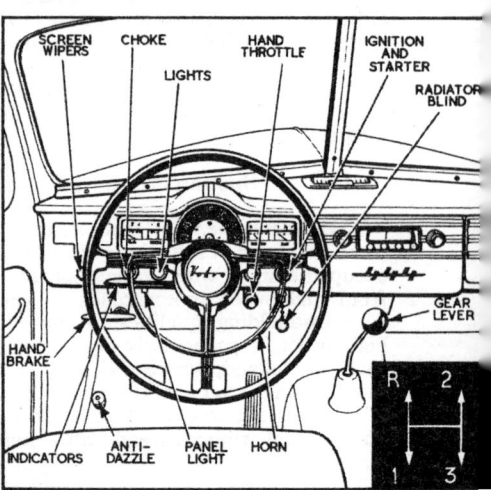

ROAD TEST .... continued

switch fitted just under the facia enables the intensity of this illumination to be varied. The instruments do not reflect in the windscreen at night.

The windscreen wipers are pivoted towards the outer edges of the two-piece, slightly V windscreen, and there is a large unwiped area of glass in the centre of the screen, although the outer portion of the screen is well cleaned. The blades are operated by induction manifold depression, and on the car tested no reservoir tank was provided, with the result that the wipers stalled as soon as the throttle was opened, a feature which seriously restricted use of the car's performance in wet weather.

The minor controls are placed below the instruments and include a combined ignition and starter switch which is arranged so that it provides current for the auxiliaries without the ignition if it is turned to the left; and for auxiliaries and ignition if turned to the right, while a further turn to the right operates the starter motor. Provision is made for the addition of a radio and heater system, although the car tested was fitted with only a radio, the control unit of which was built into the centre section of the facia. On the right there is a useful glove locker with a hinged lid, and a large shelf behind the rear seats provides useful stowage space. The main luggage locker is of moderate proportions for a car of this size, and also contains the spare wheel, although this is provided with a laminated wood cover, and thus the main compartment has a smooth floor which will not damage luggage.

Lighting

The six-volt electrical system proved to be quite satisfactory. The main beams of the double-dip head lamps give adequate range for high-speed driving at night, while in the dipped position a particularly good spread of light is provided. The horns, too, are powerful and are controlled by a D-shaped ring switch placed on the steering wheel. Flashing light direction indicators are mounted on the outside of the body high up and to the rear of the doors, in a position where they can be seen from both front and rear. They are operated from a lever on the left side of the steering column; there is a clicking device, audible from inside the car, to indicate to the driver when they are in operation. Tell-tale lights are built into the centre of the speedometer face to show which indicator is in operation, while a third warning light operates when the head lamp main beam is on.

The Volvo has a two-door body construction. The doors

The luggage locker lid is hinged from the top and fitted with a catch to keep it raised when required. The spare wheel is housed inside the main compartment but a plywood cover for it provides a flat floor for luggage. The jack and tools are also carried in the locker.

are large and it is very easy to get in and out of the front seats, while to gain access to the rear compartment it is necessary to fold forward the backs of the front seats. There is satisfactory head and leg room in both compartments and the rearward side windows, which are fixed, provide useful side vision for rear passengers. A comparatively large expanse of body panel behind the rear side windows noticeably restricts the driver's visibility in the three-quarter aft direction.

A two-point jacking system operates midway along the wheelbase, raising both wheels on one side. There is no provision for a starting handle. On the front suspension and steering 18 points require lubrication at intervals of 600 miles.

The Volvo 444 is a notable example in the 1½-litre class. It provides a very good performance and can cruise comfortably at near a more than respectable maximum speed for very long periods; it has a modest thirst and provides very comfortable transport for four persons. It is a very satisfying car to drive and covers the ground in a very willing way.

The six-volt battery is mounted in the centre of the bulkhead behind the engine, and to the left of it are the ignition coil and the light switch, which is cable-operated from the facia. A large air cleaner is located to the rear of the downdraught carburettor, and both radiator and oil filler caps are accessible. The chain running across the engine compartment operates a radiator blind.

The Motor Road Test No. 4/56

Make: Volvo
Makers: A.B. Volvo, Gothenberg, Sweden
Type: PV444K

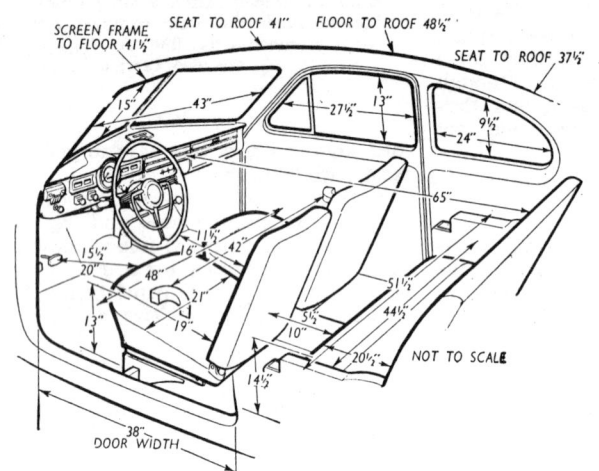

Test Data

CONDITIONS. Cool weather with very strong wind (Temperature 47-49°F., barometer 29.4—29.6 in. Hg.). Slightly damp smooth tarred road surface. Premium grade pump fuel.

INSTRUMENTS (Kilometre calibrations)
Speedometer at 30 m.p.h. .. .. 1% fast
Speedometer at 60 m.p.h. .. .. 3% fast
Distance Recorder .. .. .. accurate

MAXIMUM SPEEDS
Flying Quarter Mile
Mean of four opposite runs .. .. 75.5 m.p.h.
Best time equals .. .. 80.4 m.p.h.

Speed in gears
Max. speed in 2nd gear .. .. 59 m.p.h.
Max. speed in 1st gear .. .. 29 m.p.h.

FUEL CONSUMPTION
43.5 m.p.g. at constant 30 m.p.h.
41.5 m.p.g. at constant 40 m.p.h.
37.0 m.p.g. at constant 50 m.p.h.
32.0 m.p.g. at constant 60 m.p.h.
25.0 m.p.g. at constant 70 m.p.h.
Overall consumption for 1,050 miles, 35.3 gallons, = 29.7 m.p.g. (9.5 litres/100 km.)
Fuel tank capacity 7¾ gallons.

ACCELERATION TIMES Through Gears
0-30 m.p.h. .. .. .. 7.0 sec.
0-40 m.p.h. .. .. .. 11.1 sec.
0-50 m.p.h. .. .. .. 16.9 sec.
0-60 m.p.h. .. .. .. 28.0 sec.
0-70 m.p.h. .. .. .. 43.7 sec.
Standing Quarter Mile .. .. 23.2 sec.

ACCELERATION TIMES on Two Upper Ratios

	Top	2nd.
10-30 m.p.h.	12.0 sec.	6.9 sec.
20-40 m.p.h.	12.2 sec.	7.7 sec.
30-50 m.p.h.	12.9 sec.	10.2 sec.
40-60 m.p.h.	17.3 sec.	—
50-70 m.p.h.	26.1 sec.	

WEIGHT
Unladen kerb weight .. .. 19 cwt.
Front/rear weight distribution .. 53/47
Weight laden as tested .. .. 22½ cwt.

HILL CLIMBING (At steady speeds)
Max. gradient on top gear .. .. 1 in 11.1 (Tapley 200 lb./ton)
Max. gradient on 2nd gear .. .. 1 in 6.9 (Tapley 320 lb./ton)

BRAKES at 30 m.p.h.
0.87g retardation .. (= 34¼ ft. stopping distance) with 90 lb. pedal pressure
0.70g retardation .. (= 43 ft. stopping distance) with 50 lb. pedal pressure
0.24g retardation .. (=125 ft. stopping distance) with 25 lb. pedal pressure

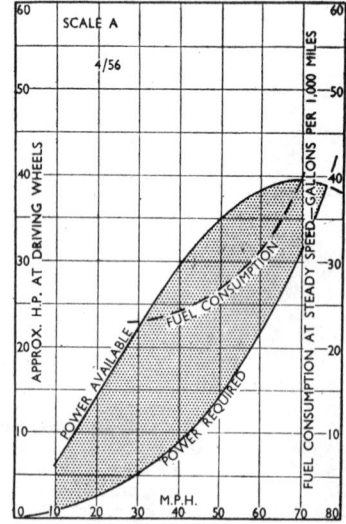

Drag at 10 m.p.h. .. .. 35 lb.
Drag at 60 m.p.h. .. .. 135 lb.
Specific Fuel Consumption when cruising at 80% of maximum speed (i.e. 60.4 m.p.h.) on level road, based on power delivered to rear wheels .. 0.69 pints/b.h.p./hr.

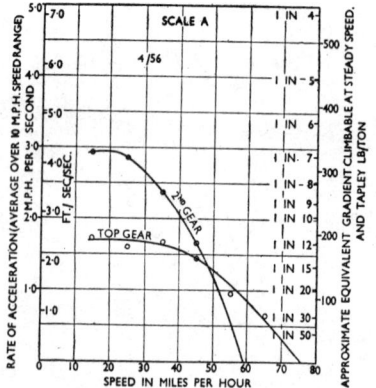

Maintenance

Sump: 6¼ pints, S.A.E. 20 Summer, S.A.E. 10 Winter. **Gearbox:** 0.9 pint, S.A.E. 90 Summer, S.A.E. 80 Winter. **Rear Axle:** 1.6 pints hypoid gear oil, S.A.E. 90 Summer, S.A.E. 80 Winter. **Steering gear:** Castrol SB, Esso gear oil 250 or Mobilube steering gear oil, etc. **Radiator:** 17 pints (2 drain taps). **Chassis Lubrication:** By grease gun every 1,250 miles to 24 points. **Ignition timing:** 2° B.T.D.C. for premium fuel. **Spark plug gap:** 0.015—0.020 in. **Contact breaker gap:** 0.015—0.020 in. **Valve timing:** I.O., 5° B.T.D.C.; I.C., 47° A.T.D.C.; E.O., 47° B.B.D.C.; E.C., 5° A.T.D.C. **Tappet clearances:** (Hot) Inlet 0.015 in. Exhaust 0.017 in. **Front wheel toe-in:** 0—⅛ in. at centre of tread. **Camber angle:** —¼° to +¾°. **Castor angle:** —¾° to +¼°. **Tyre pressures:** Front 21 lb., Rear 24 lb. **Brake fluid:** To S.A.E. Specification 70R1. **Battery:** 6 volt, 85 amp.-hr.

Ref. S/14/56.

The VOLVO 444K

A Lively, Economical and Comfortable Four-Seater from Sweden

SIMPLE in shape, the Volvo two-door saloon is an unusually comfortable four-seater. Rough roads showed off to advantage the suspension and wheel adhesion, and also the mud flaps behind the wheels, which are common in Sweden.

TO the embarrassment of those whose duty it is to test and report upon new cars, there comes occasionally a model which defies logical analysis. Such a car is the Volvo 444K, which we have recently tested on English roads over a period of two weeks. In terms of cold figures for speed, acceleration, carrying capacity and roominess, although it is a sound all-round car, it is one which, when import duty must be paid, is rather expensive for what it does. But cold figures by no means tell the whole story, for the eight members of our staff who sampled this car over significant mileages were without exception charmed by it. Not merely in what it does, but also in how it does things, this model is obviously the result of many years of careful development work on a sound basic design, by engineers who are themselves active drivers both of their own products and of cars made in other parts of Europe and America.

In general layout the Volvo is reasonably orthodox, and in dimensions it compares closely with the most popular models of this and other European countries. There is a four-cylinder engine of 1.4 litres size, with pushrod-operated overhead valves, mated to a single-plate clutch and three-speed synchromesh gearbox. Integral bodywork of rustproofed steel construction is supported by a conventional layout of coil-spring I.F.S., and by coil springs at the rear also in conjunction with a rigid axle located by torque arms. Two doors give access to four roomy and comfortable seats, and there is a separate rear luggage locker with a lid which does not lift up as high as might be wished.

In its characteristics as observed by a driver or his passengers, however, the Volvo 444K differs substantially from anything else on the British market, happily combining some virtues of both larger and smaller models. Without making the slightest pretence of accommodating five or six people, it is an unusually comfortable four-seater. It has effortless and quite high performance on the road, suggestive of an engine rather larger than that actually used, but its light controls and economy of fuel are suggestive of a rather smaller car.

Recent changes in the Volvo engine include an increase in compression ratio to 7.3/1, and a Zenith carburetter has replaced the Carter type used when we tested an earlier car in Sweden in 1950. Starting from cold remains quick, warming up performance now depends on a little use of the choke, but is good even if less remarkable than formerly, and it is still possible to use commercial-grade fuel without unduly loud pinking, although the car is happier with at least a proportion of Premium petrol in its tank. The exhaust can boom in slightly sporting fashion, especially when it is echoed by roadside walls, but this remains a smooth and flexible engine.

The clutch does its job without protest, and inconspicuously save that on the test car engagement of first gear from rest was not always silent. Controlled by a central gear lever, the three-speed gearbox has smooth and highly effective synchromesh on the upper ratios, and the unsynchronized first gear engages easily when required. The choice of ratios for a three-speed gearbox is never easy, and in this instance middle gear is much closer to the highest than to the lowest ratio, gaining thereby usefulness as an overtaking ratio for speeds up to 50 m.p.h., at the expense of rapidity of getaway from sharp corners in towns. The gearbox is by no means quiet, yet the noise produced does not prove objectionable. The good torque of the engine means that this can be driven very much as a top-gear car if desired, and despite the high second ratio it is altogether exceptional to find a lane hilly enough to require a change down into first.

As may be seen from the data page, the test car showed a two-way average speed of over 75 m.p.h. when timed over the measured ¼ mile, in adverse conditions of gale-force wind and low barometer. Over a wide range of cruising speeds, there is a pleasing surge forwards in response to any extra pressure on the accelerator pedal, 65 m.p.h. being a happy pace for long runs. Unhappily, this car suffers rather badly from wind noise if the conventional hinged ventilator windows on the doors are open, but it is possible to ignore these and to secure adequate and reasonably draught-free ventilation by winding the main windows down slightly: a detail which only two-car motorists will worry about is that the window winders work in the opposite sense to that usual on British cars.

Entry to the Volvo is at times less easy than to many other cars, for various reasons. With either of the front seats adjusted forward to suit short legs, there is limited space for swinging feet through the gap between seat cushion and front door

In Brief

Price in Britain, approx. £685, plus purchase tax, £343, equals £1,028 approx.
Capacity ... ... 1,414 c.c.
Unladen kerb weight ... 19 cwt.
Fuel consumption ... ... 29.7 m.p.g.
Maximum speed ... ... 75.5 m.p.h.
Maximum top-gear gradient 1 in 11.1
Acceleration:
 10-30 m.p.h. in top ... 12.0 sec.
 0-50 m.p.h. through gears 16.9 sec.
Gearing: 16.2 m.p.h. in top at 1,000 r.p.m.; 30.8 m.p.h. at 1,000 ft. per min. piston speed.

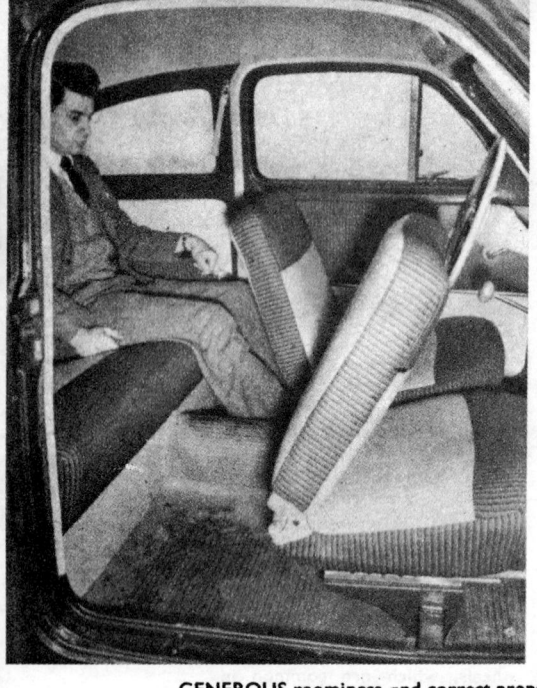

hide the offside front wing completely, but the driving position is excellent. The high position of the rear window makes reversing extremely difficult unless the correspondingly high-set mirror is used.

Use of coil springs at both front and rear may lead to expectations of an ultra-soft ride, but this is, in fact, quite a firmly sprung car by present-day standards. Whilst bumps can be felt, however, moderate unsprung weight and ample ranges of available spring travel prevent shock on either ordinary or very bad surfaces, and firm springs do not, of course, magnify waves in a road as can softer ones, nice matching of the front and rear springs giving complete freedom from pitching. Roll on corners is kept within very modest limits, the tyres do not squeal at all easily, and experiments on rough-surfaced hills in wet weather showed commendably good traction in adverse conditions.

Mechanical smoothness and precision in

The Volvo 444K - - - - -

users are likely to average 30 miles on each gallon of fuel, and substantially better economy still is not difficult to obtain. Long engine life between overhauls is also claimed, but cannot, of course, be verified within the period of a road test.

Generous in size, the brakes work with comfortable effectiveness in response to very moderate pedal pressures. Ultimate stopping power in our tests was limited to 0.87g retardation, by locking of the rear wheels, so that the braking should be even better with a full load aboard than when the car is driven two up. The handbrake, of pull-out type, is nicely placed below the instrument panel (on the left-hand-drive car tested) and the twist-to-release ratchet inspired confidence, but a rather firm pull was needed to hold the car on a steep slope.

Instruments face the driver directly on this car, their faces severely upright so that the rheostat-controlled lighting does

GENEROUS roominess and correct proportions make the Volvo rear seat unusually comfortable. Individually adjustable front seats raised fairly high above the step-down floor also give a very natural driving position.

FORWARD location of the hinges eliminates risk of the bonnet opening when the car is in motion, and allows a simple form of lock to be operated from inside the car, a safety strut preventing the bonnet blowing shut accidentally. Despite bulky heater, air cleaner and windscreen washer installations, the o.h.v. engine is reasonably accessible.

pillar, the door sill also being well above the level of the low floor. Access to the rear seat is as easy as on most two-door cars, the interior door handles being safely away from the reach of children put in the back of the car.

Concerning roominess once it has been entered, however, the Volvo is deceptive in its somewhat close-coupled appearance. The front seats are mounted in such a way as to leave a huge amount of unobstructed legroom behind and beneath them, the rear compartment being exceptionally comfortable for two people, who can really stretch their legs out and are flanked by side armrests of really convenient height and shape. The individually adjustable front seats are also high and extremely comfortable, a combined armrest and pull-handle on each door slightly reducing a passenger's wish for extra lateral support during fast cornering. Driving vision is not remarkable, the screen pillars being quite thick and the bonnet high enough to

the steering gear and other controls make this an extremely pleasant and untiring car to handle, in spite of some apparent geometrical imperfections—at speed in strong winds or on "awkward" surfaces the car needs a fair amount of conscious "driving" to keep it on its course, and the response to the steering is not by any means strictly proportional over a range of cornering speeds and radii. But the steering is so excellent mechanically that these imperfections can be ignored, and on ice the car remains stable at quite brisk cruising speeds although requiring an even lighter touch than usual on the steering wheel. Even at the lowest traffic speeds the steering is effortless, but needs only just over three turns of the wheel from lock to lock, the turning circle being conveniently small.

As has been indicated, this is an economical car to run, steady speed tests showing over 40 m.p.g. at 40 m.p.h., and over 30 m.p.g. at a mile a minute. Most

not reflect in the sloped Vee windscreen at night. The speedometer recorded distances in 1/10th kilometre units, but had no "trip" recorder, other instruments being fully adequate in number although vague in calibrations—they comprised oil pressure gauge, ammeter, radiator thermometer and fuel contents gauge. Two-blade windscreen wipers are vacuum-operated, but a reservoir of adequate size largely prevents them stopping or slowing down on hills, and a windscreen washing spray is actuated by turning the wiper knob to the left beyond its "off" position.

Interior heating is standardized on this model, with fresh air ducted to the body and windscreen interior as desired. Whilst taking several miles to become fully effective after a start from cold, the heater when working warms the whole car interior, although surprisingly efforts to ensure that warmth reaches rear-seat passengers have allowed the driver's feet to be left comparatively cold. The Volvo

- - - Contd.

SELF-SUPPORTING when open, but not lifting very far, the luggage locker lid gives access to a usefully roomy rubber-carpeted compartment.

factory at Gothenberg being beside the sea, in a country where cars are rarely garaged at night, it is reasonable to expect the chromium plating to be of good quality. An anti-theft detail is that the ignition coil is on the bulkhead, with its low-tension lead not accessible from under the front-hinged bonnet.

Unusually bright, the interior light has the three-way switch common on Continental cars, the normal setting giving "courtesy switch" operation whenever the driver's door is open, but other settings either cutting out the light or switching it on regardless of whether the door is open or shut. The headlights on the test car were outstandingly good, the 45-watt main filaments giving a shallow but long and wide beam fully adequate for fast travel on either straight or winding roads, the 40-watt dipped filaments giving a non-dazzle light rather less abruptly curtailed than the dipped beams of many Continental headlights. We have seen many lamps which are better looking in a showroom by daylight, but have seldom driven behind any better lights than these. Rather high-set stop/tail lamps which illuminate the luggage boot interior will need to be supplemented by reflectors on examples sold in Britain, but are quite separate from flashing amber direction indicators mounted centrally on the body sides. Twin horns are operated by a horn ring, which is convenient but has one "dead" section which will not sound the horns.

Points of Attraction

All in all, the Volvo made an unusually favourable impression upon all the members of our staff who drove it, as a car which would serve very varied purposes dependably, comfortably, economically and enjoyably. It is very comfortable for four passengers, reasonably happy on any road surface, quite quick over varied types of journey, and economical of fuel, concerning the quality of which it is by no means fussy. But perhaps it was an excellent driving position, mechanically smooth controls and a luxurious rear seat which most fully made up for the obvious limitations of a two-door body with deep footwells and a high bonnet line, and for various other imperfections which have been mentioned. Beyond doubt, the Volvo 444K will soon earn a loyal following if, as is anticipated, it appears on the British market in the near future.

Mechanical Specification

Engine
Cylinders	4
Bore	75 mm.
Stroke	80 mm.
Cubic capacity	1,414 c.c.
Piston area	27.4 sq. in.
Valves	Pushrod o.h.v.
Compression ratio	7.3/1
Max. power	51 b.h.p.
at	4,500 r.p.m.
Piston speed at max. b.h.p.	2,360 ft. per min.
Carburetter	Zenith 30VIG-9 downdraught
Ignition	Bosch 12-volt coil
Sparking plugs	14 mm., AC 44 Com.
Fuel pump	AC mechanical
Oil filter	By-pass

Transmission
Clutch	Borg & Beck 8-in. s.d.p.
Top gear (s/m)	4.55
2nd gear (s/m)	7.40
1st gear	14.7
Propeller shaft	Hardy Spicer divided open
Final drive	E.N.V. hypoid bevel, 9/41
Top gear m.p.h. at 1,000 r.p.m.	16.2
Top gear m.p.h. at 1,000 ft./min. piston speed	30.8

Chassis
Brakes	Wagner hydraulic
Brake drum diameter	9 in.
Friction lining area	135 sq. in.
Suspension: Front:	Coil and wishbone I.F.S., with anti-roll torsion bar.
Rear:	Coil springs and rigid axle located by torque arms.
Shock absorbers	Delco telescopic
Tyres	5.90–15, 4-ply

Steering
Steering gear	Gemmer cam and lever
Turning circle (between kerbs):	
Left	33¼ ft.
Right	34 ft.
Turns of steering wheel, lock to lock	3¼

Performance factors (at laden weight as tested)
Piston area, sq. in. per ton	24.4
Brake lining area, sq. in. per ton	120
Specific displacement, litres per ton mile	2,330

Coachwork and Equipment

Bumper height with car unladen:
Front (max.) 20 in.; (min.) 12 in.
Rear (max.) 20 in.; (min.) 12 in.
Starting handle ... No
Battery mounting ... On scuttle
Jack ... Screw type, with ratchet handle
Jacking points External, 2 on each side of body
Standard tool kit: Wheel-nut spanner, sparking-plug spanner, adjustable spanner, screwdriver, pliers and hammer (in tool bag).
Exterior lights: Two headlamps with pilot bulbs, two stop/tail lamps, number-plate lamp, two parking lamps/direction indicators.
Direction indicators: Flashing type, self-cancelling.
Windscreen wipers: Two-blade self-parking, vacuum-operated with reservoir.
Sun vizors ... Two, universally pivoted
Instruments: Speedometer with non-trip decimal distance recorder, coolant thermometer, fuel contents gauge, oil pressure gauge, ammeter.
Warning lights ... Direction indicators, headlamp main beam
Locks:
 With ignition key ... Ignition
 With other keys ... Driver's door, luggage locker
Glove lockers One on facia panel, with lid
Map pockets ... No
Parcel shelves ... Behind rear seat
Ashtrays ... One on facia, two in rear-seat armrests
Cigar lighters ... One on facia panel
Interior lights ... One above windscreen, with manual and courtesy switches
Interior heater Fresh-air type, with de-misters
Car radio ... Optional extra
Extras available: Radio, foglamp, windscreen washing sprays, loose seat covers, reversing lamps, clock, anti-dazzle mirror, external rear-view mirror.
Upholstery material ... Cloth (two-tone)
Floor covering ... Rubber mats
Exterior colours standardized: Five (black, red, grey, blue, dark blue).
Alternative body styles ... Sports two-seater on special chassis

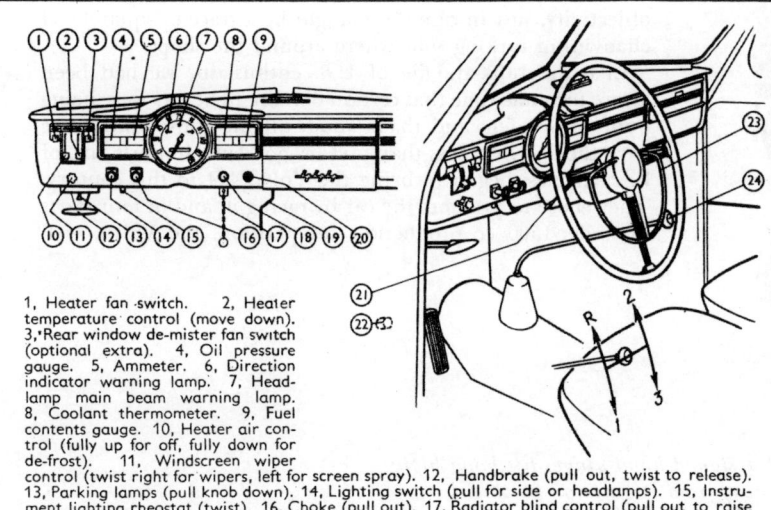

1, Heater fan switch. 2, Heater temperature control (move down). 3, Rear window de-mister fan switch (optional extra). 4, Oil pressure gauge. 5, Ammeter. 6, Direction indicator warning lamp. 7, Headlamp main beam warning lamp. 8, Coolant thermometer. 9, Fuel contents gauge. 10, Heater air control (fully up for off, fully down for de-frost). 11, Windscreen wiper control (twist right for wipers, left for screen spray). 12, Handbrake (pull out, twist to release). 13, Parking lamps (pull knob down). 14, Lighting switch (pull for side or headlamps). 15, Instrument lighting rheostat (twist). 16, Choke (pull out). 17, Radiator blind control (pull out to raise blind). 18, Cigar lighter. 19, Bonnet lock. 20, Ignition switch and starter control (twist right for ignition and starter, left for accessories only). 21, Direction indicators (turn with wheel). 22, Headlamp dip switch (press). 23, Horn ring (press). 24, Gear lever.

Swedish Invasion

SCI ROAD TEST: PV 444

Well designed hood hinges lift hood high and away from engine compartment making repair work easier. Frontal pivoting offers safety feature if hood latch is left open.

ALTHOUGH SCI's masthead bristles with Scandinavian names, Christy, Ludvigsen and Borgeson have no burning torch to carry for the land of ancestors of a few generations back. But it would be foolish to say that our first contact with a Swedish automobile wasn't marked by unusually sharp curiosity. Testing the Volvo was a special sort of adventure for us and one that put extra emphasis on objectivity, just in case there might be a trace of squarehead chauvinism lurking somewhere around the shop.

Like the rank-and-file of U.S. enthusiasts, we had been aware for some time that certain obscure light cars were being built in Sweden, and that's about all the consideration we gave them. Then, less than a year ago, Auto Imports Inc. of Los Angeles, began to bring the Volvo 444 to this country. They chose to promote the car by racing it, and in four starts in under-1500 cc production competition, the Volvo, well

Looking at the Volvo head on from a low angle gives the car a high truck appearance. Actually height is 61.5 ins.

driven by Ron Pearson, easily ran away from the pack and won four firsts.

Here was a car really to be contended with. The citizenry was impressed and so were we. From its first appearance, the Volvo had set an entirely new standard for competition in its class. But it wasn't until we test-drove the Volvo briefly for SCI's July light-car survey that we began to realize how much automobile reposed beneath this car's disarming exterior. Now, after extended road testing of the $1995 Volvo, we're prepared to state that this is the hottest light car you can touch within a thousand dollars of its price at this time. It may not win any *concours d'elegance but on all grounds other than—in some case—esthetic ones, it can nail its competitors to the wall.* Let's start with speed.

Our second Volvo, the one submitted to full-scale test, had just been driven back from the CSCC races at Bakersfield, California, and had over 3000 miles on its odometer. It had finished second there (Volvo's fifth race on the Coast) to an expensively-tuned, much more costly car. In spite of its accumulated mileage, this Volvo still felt tight and owners have assured us that these cars don't begin to really loosen up until after the first 10,000 miles or so. Yet with a mere 1.5-mile approach to our timing traps, we were able to clock a true 94.8 one way and 94.1 as a two-way average. This is nice going for a one-ton car with a displacement of just 1414 cc or 87 cu. ins.

It's remarkable. The Volvo is faster than the not fully broken-in Alfa Giulietta Sprint tested by SCI and its top speed is about equal to that of a typical MG "A" or Porsche 1500 Speedster. Kick that one around at your next bench race.

To get up to almost 95 mph in a pretty short distance, the Volvo has to be a good accelerating car. Actually, it's in a

Interior view shows the roominess of driver's niche, the attractive finish, and the advantageous stick shift.

In profile, the Volvo might be a shortened '40 Ford, but there the similarity ends. The quality of the workmanship is unusually high, and the paint is rubbed to a glassy finish.

RATING FACTORS:

Bhp per cu. in.	.81
Bhp per sq. in. piston area	2.56
Torque (lb-ft) per cu. in.	.87
Pounds per bhp — test car	30.6
Piston speed @ 60 mph	1910 fpm
Piston speed @ max. bhp	2890 fpm
Brake lining area per ton (test car)	108.5

SPEEDOMETER CORRECTION:

Indicated	Actual
30	28.0
40	38.0
50	48.5
60	58.5
70	69.3
80	80.0

The twin SU carburetor fed engine with the rocker assembly exposed. Tips of the rocker arms are case hardened and springs pack 172 lb. pressure. Battery (not seen) is a six volt unit.

Entering SCI test curve at 50 mph, car rode flat and steady. Some lean can be detected, however, in photo. Good cornering comes from its almost equal weight distribution, with 2 aboard.

Taking leave of the curve, the 444's attitude has understeer effect, tail out, nose into turn. With a back seat passenger or some trunk luggage, 50/50 distribution can be hit on nose.

Instrument panel consists of heat and vent controls at left and gauges at right. In first shipments, gauges had metric and Swedish markings. Latest exports have English readings.

class by itself. For our own information we made up a single chart containing the superimposed acceleration curves of all the cars covered in SCI's light-car survey. We found that the Volvo's curve bears no resemblance to all the others. It is steeper and higher and is more of a sports car than a typical light car.

The factory claims that the Volvo should move from zero to 60 mph in 19.6 seconds. Our capsule-test car, a well flogged race veteran and sales demonstrator, would yield no better than 21.5. But the second car sprang effortlessly from rest to an actual corrected 60 in just 17.3. In the eight car field the Volvo stands out as a real high performance vehicle. It represents a blending of good sports car qualities with the best of the light car's virtues.

There's a further combination in the Volvo of characteristics that usually are mutually exclusive and this cannot be expressed more neatly than it was in a test report in THE AUTOCAR (London). The Volvo 444, it stated, "combines the general comfort and convenience associated with a vehicle of transatlantic (Detroit) design with the handling qualities that are expected of a thoroughbred Continental car. Added to these qualities is a third, that of good general and detail finish and sound engineering."

The Volvo definitely reflects strong Detroit influence in its body design, roominess and generous interior appointments. Also, it's equipped with a three-speed transmission, a concession to convenience as opposed to sporting requirements. This is or is not an advantage, depending upon your point of view. Had the Volvo been designed with heavier emphasis on sheer performance, a four-speed box certainly would have been specified. The car's acceleration times could be greatly improved by more closely related torque-multiplication factors, and corners could be approached at higher speeds if there was a cog between Second and High to drop into. But even so, the Volvo remains practically unbeatable in its displacement class and the star performer in its price bracket.

The floor-shift transmission is a delight to use because of the precision of its action and the infallibility of the synchromesh provided on the two top gears. Unlike most synchro boxes, with which it's necessary to allow a moment for gear speeds to become equallized during a shift, the Volvo's cogs can be popped back and forth in the upper ratios just as fast as you please, and there is never a trace of noise or clashing during this operation. The useful range of Second is remarkably broad and it's possible to wind right on out to 60 mph if you want to push the engine. Low is a noisy gear to get into a standstill unless you first pop the lever in High and then into Low, which then produces a completely silent change. Silent downshifting to Low is an exercise that calls for skillful double-clutch and throttle work.

While the Volvo can be driven just like any stick-shift American car, its road manners are drastically different. It can be paid no higher compliment than to say that it handles very much like a good Italian road machine. To understand why the Volvo behaves as it does, let's take a look at the market it originally was designed for.

Sweden has its share of primitive roads, plus a very cool climate with long, frigid winters, almost no spring at all, and short, tepid summers. In the northern part of the country the ground may be covered with snow for six or seven months out of the year. So the Volvo is a car designed, among other things, to cope as well as possible with road surfaces that are rough, rutted, frozen and slippery a great deal of the time.

The result is a vehicle with the go-anywhere abilities of a GMC Recon wagon plus authentic sports-car roadholding. It can be taken through turns at surprisingly high speeds without apparent body roll, without a murmer from the tires and with a minimum of effort on the part of the driver. It is a safe-cornering car, and feels like one. It can be drifted with confidence if you want to push it that hard and the rear tires can be slipped deliberately, although the Volvo is fundamentally a tracker, rather than a slider, in the turns. It has a very slight amount of understeer.

Much of the Volvo's good cornering stance is derived from its weight distribution. Unloaded, it is slightly nose-heavy. With two in the front seats, weight on the front and rear axles closely approaches equality. With a rear seat passenger or two or with some ballast in the luggage compartment, 50-50 can be hit on the nose.

Front wheel suspension is conveniently independent by coil springs and unequal-length wishbones. The solid rear axle also uses coil springs and is located fore and aft by a pair of husky, hat-section torque arms. Stiff anti-roll torsion bars are fitted at both front and rear. It's a well-knit, heavy-duty suspension layout, more ruggedly substantial than you'd expect to find on a light car.

Precise control of the car is abetted by ZF worm and two-stud lever (cam and lever) steering, as also fitted to the early 300SL's and to current Alfa Romeos. The just-over-three-turns from lock to lock provides as quick response as you are likely to need or want in a touring machine. There is no play in this steering and it is very light, even during parking

TEST CAR:
Volvo 444 sedan, "sports" engine. Over 3000 miles on odometer

TOP SPEED: (1½ mile approach to ¼ mile trap)
Two-way average .... 94.1 mph
Fastest one-way run .... 94.8 mph

ACCELERATION:
From zero to	Seconds
30 mph	4.9
40 mph	7.5
50 mph	11.6
60 mph	17.3
70 mph	22.9
80 mph	47.2
Standing ¼ mile	21.2
Speed at end of quarter	68 mph
Standing mile	57.4 secs.
Standing mile, average speed	54.2 mph

SPEED RANGES IN GEARS:
I .... Zero to 28 mph
II .... 9 to 56 mph
III .... 19 to 94 mph

FUEL CONSUMPTION:
Hard driving .... 21 mpg
Average driving (under 60 mph) .... 29 mpg
In heavy traffic .... 23 mpg

BRAKING EFFICIENCY:
(5 successive emergency stops from 60 mph, just short of locking wheels):

	Percent
1st stop	70
2nd	70
3rd	70
4th	70
5th	68

POWER UNIT:
Type	In-line four (three main bearings)
Valve arrangement	Pushrod ohv
Bore & Stroke (Engl. & Met.)	2.95x3.15 ins. 75x80 mm
Bore/Stroke Ratio	1/1.07
Displacement (Engl. & Met.)	86.6 cu. ins. 1414 cc.
Compression Ratio	7.8 to one
Carburetion by	Two S.U.
Max. bhp @ rpm	70 at 5500
Max. Torque @ rpm	75.2 lbs.-ft. at 3000

CHASSIS:
Wheelbase:	102.5 ins.
Front Tread	51 ins.
Rear Tread	51.5 ins.
Suspension, front	Independent, by coil springs and unequal-length A. arms. Anti-roll torsion bar
Suspension, rear	Coil springs, radius arms. Anti roll torsion bar
Shock absorbers	Double-acting hydraulic
Steering type	ZF cam and two-stud lever
Steering wheel turns L to L	3.25
Turning diameter	33.5 ft.
Brake type	Hydraulic, leading and trailing shoes F & R

GENERAL:
Length	177 ins.
Width	62.5 ins.
Height	61.5 ins.
Weight, test car	2140 with full fuel tank
Weight distribution, F/R	52.4/47.6
Weight distribution, F/R, with driver	51.8/48.2
Fuel capacity — U.S. gallons	9.5
Ground clearance	8 ins.

maneuvers, yet steering feel is good at all speeds. No road shock is telegraphed up the steering column.

The Volvo has an honest, unaffected character and the job it undertakes it does with craftmanlike thoroughness. The ride and general tautness of the chassis, for example, never let you down. The car feels solid and secure at 40 mph and feels no different at top speed. There's no body shake or driveline vibration to give you the feeling at high velocities that things are beginning to come unglued.

The chassis' shock-absorbing ability is up to good modern practice on average road surfaces. But when you hit bad roads in the Volvo you find its reaction hard to believe. You can charge at 50 mph through horrible ruts and chuckholes that would shatter the running gear and occupants of most cars. Yet you feel scarcely a ripple in the Volvo, and this statement is meant *literally*, not figuratively.

Scarcely any engine vibration is felt inside the car at any speed and the engine is quiet except at idle, when it emits a pleasant tic-tic that sounds like a chronometric tach digesting rpm. As soon as a load is put on the engine, its sound changes to a quiet hum and it remains smooth up to the peak-rev range, when the rocker box naturally begins to make itself heard.

Extreme reliability is claimed for this short-stroke four, and the factory, usually conservative in its statements, says that it is not unusual for a Volvo to go for more than 120,000 miles before the first rebore. Many pains have been taken to build this durability into the engine, including hardened surfaces on all bearings of the three-main crankshaft, porous-chrome piston rings, lead-bronze bearing inserts, case-hardened rocker arms, and spectacularly machined cam followers.

The engine's designer obviously has a passion for control of internal temperatures, and so much water jacketing is provided around the cylinders and in the head that the fan is almost entirely unnecessary. The cooling of plugs, combustion chambers, exhaust valves and cylinder bores should never be a problem. We understand that large numbers of Volvos are sold in North Africa and the Near East, and that in the most torrid climates they are never known to overheat, which is easy to believe.

Such is the extent of this emphasis on engine cooling that a warmup blind is provided in front of the radiator, operated from the instrument panel by means of a chain control. This arrangement lacks the automatic convenience of a thermostat, but it does permit the driver to select any operating temperature that he may choose. In an atmospheric temperature of 70 degrees F. we made a 40-mile run with the blind all the way down. This included a final burst of 90 mph for several miles. Upon stopping, you could rest your hand on the top tank of the radiator and feel no discomfort.

For rapid warmup the blind can be raised all the way, then lowered partially to maintain a reasonable operating temperature. It's all too easy, we found, to forget to lower the blind from fully-raised and then note that the needle on the water-temperature gauge has shot all the way out of sight. This happened to us on two occasions, proving what slaves we've become to automatic controls. But we lowered the blind, kept going at a moderate speed, and within a minute, the needle was back in the permissible range.

The performance factors (see tables) relating to engine efficiency are very good when judged by light car and even by production sports car standards. One of the main contributing elements to this good performance is the excellent breathing of the Volvo sports engine, achieved by means of dual side-draft carbs, over-sized valves and ports, the intake ports being polished, beefed-up valve springs, and a cam that packs a mildly warm grind. The lag you'd expect in bottom-gear getaway from rest with an engine of this size is there all right, but it's briefer than you might expect, and the pistons begin pumping very effectively at low revs. Most of the engine's components likely to require replacement in time are popular U.S., English, and German parts, widely available here. A six-volt electrical system is used.

Returning to the Volvo 444, this combination sports-family vehicle loses points on very few grounds. Its brakes appear to be its most serious mechanical shortcoming. Their lining area per ton factor is rather average and their performance is not up to the high standard set by all of the car's other mechanical elements. This definitely does not mean that the brakes are bad; in fact, they're a shade better than the Detroit average. But they do fade noticeably and during our ten-stop test registered a big decrease in efficiency, although still retaining highly effective stopping power.

The Volvo's body matches or surpasses Detroit standards of finish and solid construction; it forms a unit with the frame and is Bonderized for rust-resistance. Its paint is glass-smooth. It is one of the extremely rare bodies in the light car field that the average U.S. motorist can enter for the first time and feel quite at home in. It has all the familiar conveniences: cigarette lighter, three ash receivers, conventional (and legal) turn indicators, dome lights, seats that are well, well off the floorboards. It's a five passenger car and the seats can be folded to make a roomy bed. The rake of the front seat-backs is adjustable.

Like the early VW's, the first shipments of Volvos lacked some export refinements. Recent shipments are finished to a fine turn. Interiors, instrument calibration and labelling, and bumpers are all now competitive with Detroit. It appears obvious that Sweden is willing to do its best for the Yankee dollar.

In a nutshell, the Volvo 444 is a short wheelbase car, but is not a small car. It has the fuel economy and nimbleness of a light car, but is free of the performance limitations and claustrophobic disadvantages of many of these. It has a great deal of the zip and agility of a light production sports car, yet it's a family-sized machine. It's a skillful combination of good things from both sides of the Atlantic and as such is in a class apart.

Griff Borgeson

SEE VOLVO • DRIVE VOLVO • YOU WILL BUY **VOLVO**

85 HP SPORTS ENGINE
95 MILES PER HOUR TOP SPEED
OVER 30 MILES PER GALLON
Ask about the Volvo Overseas Delivery Plan

A PRODUCT OF SUPERB SWEDISH ENGINEERING

VOLVO DISTRIBUTING, INC.	SWEDISH MOTOR IMPORT, INC.	AUTO IMPORTS, INC.
452 HUDSON TERRACE	3301 WEST 12th STREET	13517 VENTURA BLVD.
ENGLEWOOD CLIFFS, NEW JERSEY	HOUSTON 24, TEXAS	SHERMAN OAKS, CALIF.

IN CANADA: AUTO IMPORTS (SWEDISH) LTD., 48 CROCKFORD BLVD., SCARBOROUGH (TORONTO) • 1350 EAST GEORGIA ST., VANCOUVER 6

Unlike the standard model, the California saloon has a badge bar crossing the grille above the overriders, and there are separate sidelights. The two-door saloon body is well sealed and comfortably upholstered

Autocar ROAD TESTS

No. 1608

VOLVO PV444
CALIFORNIA SALOON

ALTHOUGH the Volvo company has been exporting a wide range of vehicles to more than 40 countries for some years, the cars are not well known to most motorists outside Sweden, their home country. During the last year or so, however, the success of an English driver at the wheel of a Volvo in American races has helped to swell the demand in that country for this particularly sturdy car, and now the volume of exports is increasing steadily. It is centred on the PV444 California, which is similar to the standard model, with the exception that the engine is tuned to develop 70 b.h.p., with the aid of twin S.U. carburettors, instead of the standard 51 b.h.p.

This model will have its first formal showing to the public here at the London Show in October, but a full test has been carried out in Sweden on a car made available by the factory.

To fit this car into the motoring scene it is necessary to know a little of the manufacturing background. Volvo policy, constant since the introduction of the post-war saloon, has aimed at detail improvement and simplification of service and repair work, rather than radical change in specification. The result has been that in Sweden decarbonization, for example, is carried out in the minimum of time simply by substitution of the cylinder head, and special replacement assemblies have been evolved that enable accident damage to be repaired quickly and cheaply.

So efficient are the present repair facilities that buyers in Sweden are covered by the factory for all accident damage over and above third-party insurance cover and the first 200 Kroner (nearly £14) which must be paid by the owner. Thus, provided he has not imbibed unwisely, if a driver skids on the winter ice into a tree, he pays only £14 towards the total cost of repairs.

In the California the policy of simplifying servicing has resulted in a useful refinement, particularly for overseas buyers, in the reduction of greasing points from 16 to four by the use of impregnated plastic bearings. Those in the "out-back" will also appreciate the design of the seats which permits conversion into sleeping room for two.

The specification is tabulated in detail in the usual manner of Road Tests, but briefly the California seats four people in comfort, and its four-cylinder, 1,414 c.c. o.h.v. engine will propel the car at speeds up to a true 90 m.p.h. in slightly favourable conditions. In general conception it has many similarities to other European models, resulting from the requirements of Swedish drivers who have to travel long distances on narrow roads which are often indifferently surfaced, or even constructed of unsurfaced dirt.

Apart from a face lift at the front, the general shape remains similar to that used for some years, and there is no immediate intention to make radical alteration. The

The sturdy central gear lever operates a three-speed box. All the instruments can be seen easily through the upper half of the steering wheel. The heater controls are at the left of the facia. There are separate front seats, and armrests on the doors

28

VOLVO PV444
California Saloon . . .

Twin S.U. carburettors with separate air cleaners are used on the 70 h.p. model. Access is straightforward to all items that must be checked regularly. The powerful heater unit is attached to the bulkhead on the left side, and the oil filter can be seen beside the left front corner of the engine

lines are unassuming, but the model's rugged construction rose-tints the average owner's spectacles until he looks upon the car with the appreciation and affection given to a favourite pet of the household. And to the amiability of the export Volvo is added high performance with safe handling characteristics.

Starting is immediate, and in temperate weather little or no use need be made of the choke. Clutch take-up is smooth, and gear selection is effected easily with the stoutly constructed central gear lever. There are three forward speeds, with synchromesh on the upper two.

Sheer Performance

When maximum acceleration is required the gear lever can be slammed through without fear of beating the synchromesh, and some indication of the car's liveliness is to be found in the impartial data revealed by the stopwatch. From a standing start 30 m.p.h. is reached in 5.5sec—a modest figure resulting partly from the necessity of a change from first to second at a maximum of 26 m.p.h. But thereafter the car gets away remarkably well, having regard to the engine size. The speed rises to 60 m.p.h. in less than 20sec and 70 in fractionally over the half-minute, and the standing quarter-mile is covered in a creditable 20.9sec.

The tuning that produces this performance has not seriously impaired flexibility in traffic. On top gear the engine pulls smoothly from 16 m.p.h. There is some jerkiness in town driving, which seems to result from the throttle mechanism rather than engine temperament. At slow speeds particularly, it is not always easy to open the throttle smoothly, and so the meticulous driver may wish to ease the clutch a little when wanting to surge forward evenly with the traffic queue.

Cruising speed is potentially high, as the engine revs freely without exhibiting distress and without any roughness. Certainly speeds in the seventies can be used indefinitely, and 80 m.p.h. or more can be sustained when required. A small air cleaner is fitted to each carburettor, but they fail to deaden induction roar. When the engine is pulling hard the effort is audible, the mechanical noise level being of the sports type, rather than that of the ordinary family saloon. Conversation can be conducted in normal tones, however, the engine noise being of a kind that makes itself noticed without actually constituting a nuisance.

First gear on any family car should be sufficiently low to permit towing a trailer or the car of a friend in distress,

Backrests of the front seats fold forward to permit access to the rear passenger compartment. The propeller-shaft tunnel is deep, but this is of little consequence as the car is primarily a four-seater. When required the seats can be rearranged to form sleeping accommodation for two people

which means, as a corollary, that there should be three other forward gears if the car has a high performance. As the Volvo lacks a fourth gear it suffers accordingly, but the middle and top ratios are well chosen. A maximum of 61 m.p.h. can be reached in second,' which is considerably higher than the second gear maximum of the majority of comparable cars with three-speed boxes. Yet the gap between first and second is not so wide as to be inconvenient when coupled with the characteristics of this engine, for in second gear it provides an ample amount of torque at low speeds.

The emphasis given to the sheer performance of this California model is justified, but it leads immediately to consideration of the brakes, as the most important single factor governing use of the power. Sweden has winding roads that are sparsely populated with traffic, and downhill slopes alternate with uphill almost to the exclusion of flat stretches. The brakes evidently have been designed for

In its latest form the Volvo has a rectangular grille. The wings are detachable for simple replacement in the event of damage. The windscreen is divided centrally, but visibility is not seriously affected

these conditions, and even under arduous test driving no sign of fade occurred. The pedal pressure for maximum retardation is modest, and the brakes continue to act without any sort of deterioration no matter how the car is driven in normal road work.

The suspension strikes a particularly good balance between the resilience needed for relatively high speed on rough surfaces and the demands of the average family. It is firm, and a bump is felt as a bump; yet the wheels retain contact with the road almost regardless of speed and surface. The car is inherently stable, and the driver finds that long distances can be covered without fatigue because conditions under-wheel are gently indicated all the time through the suspension and steering. Control is in the driver's hands, not in the characteristics of the car.

At speed on Sweden's dirt roads there is no wheel patter, and the passengers enjoy a smooth ride. On good surfaced bends taken at high speed there is a moderate amount of roll. The steering has only 2¾ turns from lock to lock, which aids accurate control.

Some shock is felt at the wheel, but there is no undesirable kick-back. A mild degree of understeer is built in to the steering-suspension system, which will appeal to the majority of drivers: the car, in other words, is unlikely to catch one napping. The handling characteristics generally are of a high standard for a family saloon.

Although the steering column is not adjustable, the driving position suits drivers of almost any height. When the car was inspected at the factory it was observed that a Volvo staff member 6ft 3in tall was completely comfortable behind the wheel.

The speedometer, calibrated in m.p.h. for the British and American markets, is directly in front of the driver, seen through the unobstructed upper half of the steering wheel. It is reasonably accurate. The fuel and oil pressure gauges, water thermometer and fuel level indicator flank the speedometer, where they also can be read easily. Minor control knobs are arranged in a neat row below the instruments, and can be reached without bending forward. At the passenger's end of the facia there is a glove compartment, with lid but not lockable.

Door locks and window wind mechanisms are orthodox, and there are swivelling ventilating windows at the front. Windscreen and rear window are both divided. Forward visibility is reasonably good in spite of the divided screen, but rearward vision is poor. The driving mirror reveals a wide but shallow strip of the road behind, and large blind areas at the rear corners are a nuisance in reversing manœuvres.

There are no reflections in the screen from the instrument lighting, which is rheostat-controlled, and the wipers sweep wide arcs. The head lamps give adequate illumination at night.

Sealing and Heating

Swedish climatic and road conditions are reflected in the attention paid to heating and dust sealing. There are controls for air temperature and distribution, and switches for two two-speed fans, the second of which is an optionally extra fitting at the rear. When going at full strength, the heating system bids to rival a blast furnace. Quick warming up of the cooling system is aided by the radiator blind fitted as standard, and controlled from within the car. On dry dirt roads, which throw up a thick, fine dust, the body sealing was found to be perfect. The good sealing was indicated also by the need to slam a door sharply when all other orifices were shut.

Entry to the front compartment is easy, and the two doors are forward hinged for safety. Access to the rear requires one or both of the backrests of the front seats to be folded down. It presents no serious difficulty, as the doors are wide and swing through a substantial arc. All the seats are upholstered in hard-wearing cloth of attractive appearance, and are comfortable. There is plenty of headroom at front and rear, and with the front seats adjusted for people of average height there is no lack of leg room for the rear passengers.

The spare wheel is mounted vertically on the right-hand side of the compartment, and the tools are housed in a recess hidden behind the lip of the locker opening. The amount of room available is considerable, but the number of rectangular suitcases that may be carried is limited by the shape of the locker as a whole

VOLVO PV444 California Saloon . . .

Luggage accommodation is fair, the tapering tail giving a shape to the locker which limits the number of suitcases of conventional type which can be carried. The spare wheel is mounted vertically at the side where it cannot damage luggage, and the tools lie in a shallow trough at the rear of the locker floor where they are very convenient to reach. Jacking sockets attached to the frame are provided for each wheel.

The bonnet is hinged at the front, and locked by a lever inside the car. The lever is reasonably accessible, but must be used to re-secure the bonnet when it is closed. Accessibility to the engine compartment is good, water filler, dipstick, battery and carburettors all being easy to reach.

The handbook is available in English, and its contents reflect the care taken to simplify home maintenance. The specification is detailed, including even the correct torque for wheel nut tightening. The best method of connecting a tow-rope is illustrated, and there are detailed instructions for removing stains from upholstery and paintwork.

When considering the Volvo as a whole a few points of criticism arise. The sports engine is a little noisy, the general appearance somewhat homely, and rearward visibility below average. In other respects the car is thoroughly sound, giving a strong suggestion of inherent honesty. The performance is exceptionally high for a car of this type and the chassis permits full advantage to be taken of that. Long journeys leave all occupants feeling fresh.

The Volvo is of Swedish design and of 80 per cent Swedish manufacture. Of the remaining components about 17 per cent, including brakes, wheels and carburettors, are made in England.

VOLVO PV444 CALIFORNIA SALOON

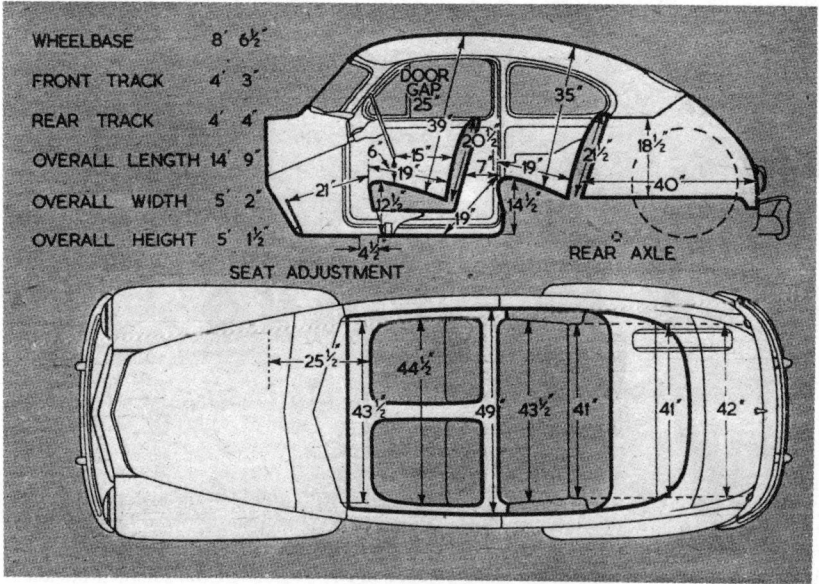

Measurements in these ⅛in to 1ft scale body diagrams are taken with the driving seat in the central position of fore and aft adjustment and with the seat cushions uncompressed

SPECIFICATION

ACCELERATION: from constant speeds.
Speed Range, Gear Ratios and Time in sec.

M.P.H.	4.56 to 1	7.40 to 1	14.7 to 1
10—30	..	—	6.6
20—40	12.2	6.5	—
30—50	13.5	7.1	—
40—60	14.7	—	—
50—70	18.5	—	—

From rest through gears to:

M.P.H.	sec.
30	5.5
50	12.7
60	19.6
70	30.2

Standing quarter-mile, 20.9 sec.

SPEEDS ON GEARS:

Gear	M.P.H. (normal and max.)	K.P.H. (normal and max.)
Top (mean)	85	136.8
(best)	90	144.8
2nd	44—61	70.8—98.2
1st	20—26	32.2—41.8

SPEEDOMETER CORRECTION: M.P.H.

Car speedometer:	10	20	30	40	50	60	70	80	84	94
True speed:	10	20	30	40	49	58	66	77	80	90

TRACTIVE RESISTANCE: 37.4 lb per ton at 10 M.P.H.

TRACTIVE EFFORT:

	Pull (lb per ton)	Equivalent Gradient
Top	225.1	1 in 9.9
Second	336.2	1 in 6.6

BRAKES:

Efficiency	Pedal Pressure (lb)
34.5 per cent	25
74.0 per cent	50
79.8 per cent	75

FUEL CONSUMPTION:
33 m.p.g. overall for 560 miles (8.56 litres per 100 km).
Approximate normal range 28.5—39.5 m.p.g. (9.9—7.15 litres per 100 km).
Fuel, first grade.

WEATHER: Dry, stiff head/following breeze.
Air temperature: 54 deg F.
Acceleration figures are the means of several runs in opposite directions.
Tractive effort and resistance obtained by Tapley meter.

DATA

PRICE (basic), with two-door saloon body, £740 approx.
British purchase tax, £360 approx.
Total (in Great Britain), £1,100 approx.
Heater standard.

ENGINE: Capacity: 1,414 c.c. (86.65 cu in).
Number of cylinders: 4.
Bore and stroke: 75×80 mm (2.95×3.15in).
Valve gear: o.h.v. pushrods.
Compression ratio: 7.8 to 1.
B.H.P.: 70 at 5,500 r.p.m. (B.H.P. per ton laden 63.6).
Torque: 75.9 lb ft at 3,000 r.p.m.
M.P.H. per 1,000 r.p.m. on top gear, 16.2.

WEIGHT (with 5 gals fuel): 19 cwt (2,128 lb).
Weight distribution (per cent): F, 51.5; R, 48.5.
Laden as tested: 22 cwt (2,464 lb).
Lb per c.c. (laden): 1.74.

BRAKES: Type: Leading and trailing.
Method of operation: hydraulic.
Drum dimensions: F, 9in diameter; 2in wide. R, 9in diameter; 2in wide.
Lining area: F, 67½ sq in. R, 67½ sq in (118.8 sq in per ton laden).

TYRES: 5.90—15in tubeless.
Pressures (lb per sq in): F, 20; R, 22 (normal).

TANK CAPACITY: 7¼ Imperial gallons.
Oil sump, 6½ pints.
Cooling system, 14 pints (including heater).

TURNING CIRCLE: 33ft 6in (L and R).
Steering wheel turns (lock to lock): 2¾.

DIMENSIONS: Wheelbase: 8ft 6½in.
Track: F, 4ft 3in; R, 4ft 4in.
Length (overall): 14ft 9in.
Height: 5ft 1½in.
Width: 5ft 2in.
Ground clearance: 8in.
Frontal area: 20 sq ft (approximately).

ELECTRICAL SYSTEM: 6-volt; 85 ampère-hour battery.
Head lights: single dip; 45-40 watt bulbs.

SUSPENSION: Front, independent, coil springs; anti-roll bar. Rear, coil springs.

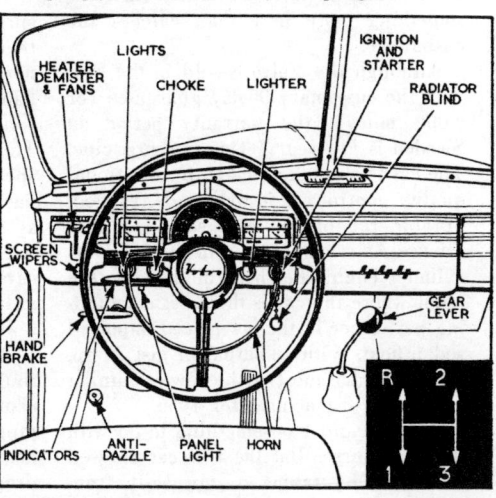

Performance and quality offset the lack of styling advances on the Volvo. The grille bars, oddly, have a galvanized coating.

Road test... SWEDEN'S HOT VOLVO

There's more here than meets the eye—it's an import with quality plus acceleration

WHEN looking over the Volvo, little evidence can be found of any postwar features that have come to be more or less standard with the U.S. car, as well as a good many European vehicles. The styling strongly resembles that of the earlier Ford or Mercury, especially in a profile view. In support of this impression is the divided windshield, the long lever of the floor-mounted shift that vanished from the scene some two decades ago, the broadcloth upholstery.

Yet the Volvo, which is being imported from Sweden in substantial quantities, is selling remarkably well. It is a solid favorite with a rapidly growing group of owners and drivers. Obviously, there is something more here than catches the casual eye.

Although the Volvo is sold in the U.S. with the customary 90-day guarantee (or 4,000 miles), the warranty period in Sweden is *five years!* This is a first clue. The car is manufactured with exceptional quality control; it is well-made, rugged, durable, reliable. So much so, in fact, that for the American market the tough four-cylinder engine has been modified to the point where the car is the most outstanding performer in the low-priced imported sedan field. With an output of just 70 hp, it'll even edge most of the few remaining U.S. non-V-8 machines for 1956.

Light weight has something to do with this, of course. But the test car showed its worth by taking a downshift from third into second at 90 mph without excessive protest. It is at higher rpm that the car reveals its unusual performance; in the lower ranges it is deceptively docile. The best shift points when going hard, incidentally, is from low at 35 mph and holding second until about 68 mph. The under-five-seconds time to 30 mph is better than some of the current V-8's out of Detroit can do.

The engine sound is somewhat like that of a Triumph or an MG-A. The hotter recently imported Volvos have a new valve layout, 14 mm plugs instead of 10 mm, and a more radical cam, which makes quite a difference.

Volvo's suspension design includes coil springs on all four wheels. The ride, therefore, is a good deal softer than the normal European car offers. On the other hand, the sacrifice in handling to achieve soft springing has not been too great. The Volvo will easily negotiate corners at speed; This, coupled with its hot engine, has made it a car to beat in sedan events at recent sports car road races. Its lap times have been equal to many sport cars—a remarkable feat considering it's a large 102-inch wheelbase sedan with a substantial frontal area.

A further point likely to develop enthusiasm among owners is the good fuel economy. The test car averaged nearly 30 mpg during normal city driving; when acceleration tests were conducted on a drag strip, a day of floor-boarding the throttle still yielded an average of 22 mpg.

There is good legroom both front and rear in the Volvo, but the forward bucket seats limit comfortable seating there to two persons. The rear bench-type seat could handle three. Also, the rear side windows are fixed, cannot be rolled down, which is a distinct disadvantage. As far as instruments are concerned, the Volvo has the normal variety in the conventional arrangement. Nor is there anything peculiarly foreign about the layout. The steering wheel is high, but not uncomfortable, the pedals are not inconveniently arranged, as is often found in an European-made car.

The Volvo is priced at $1995, with most items of equipment standard (except radio). This brings it down into the low-priced field, both imported and domestic. Although the company makes other body types, including a sports car, this is the only car currently being imported under the name. However, another is expected within a few months.

The low-priced imported car field is a pretty competitive bracket. In order to make any headway in it, a product has to be better than good. The Volvo lacks the advantages of contemporary styling. But its better construction, superior acceleration for its class matched with fuel economy and durability, are fashionable features, too. For those who appreciate such sturdy qualities, the Volvo has 'em.

Four-cylinder engine has been highly modified for U.S. market, but extremely rugged design permits greater output without reducing its life or reliability on the road.

PERFORMANCE

ACCELERATION: 0-30 mph in 4.8 seconds, 0-45 mph in 11.2, and 0-60 mph in 17.2
TOP SPEED: 90-92 mph
GAS MILEAGE: 29 mpg in normal driving
SPEEDOMETER CORRECTIONS: Indicated 30, 45 and 60 mph are actual 28, 42 and 57 mph

SPECIFICATIONS

TEST CAR: 1956 Volvo PV 44
BODY TYPE: two-door sedan
BASIC PRICE: $1995
ENGINE TYPE: ohv four cylinders
DISPLACEMENT: 86.6 cubic inches
COMPRESSION RATIO: 7.9-to-1
HORSEPOWER: 70 @ 5500 rpm
TORQUE: 75.9 lb. ft. @ 3000 rpm
DIMENSIONS: Overall length 177 inches, width 62½, height 61½, wheelbase 102½, front tread 51, rear tread 51½
WET WEIGHT: 2125 lbs.
TRANSMISSION: Conventional, three speeds forward

Interior is bare of plastics, either in upholstery or trim. The legroom on both sides of floor-mounted gear lever is good.

Sloping rear on Volvo body makes it one of few remaining fastbacks still in production. The rear side windows are fixed in position, cannot be cranked down.

33

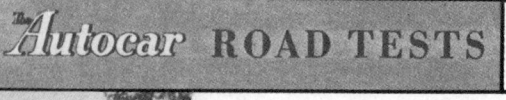

Volvo PV444

WITH SPORTS ENGINE

Functional rather than elegant lines distinguish this Scandinavian, rarely seen in Britain. At high speed there is little wind roar

SINCE the last full Road Test of the Swedish Volvo PV444 saloon appeared in *The Autocar* of 7 September 1956, a sports version has been marketed. This, while generally similar to the earlier car, has a different engine which provides remarkably good acceleration and maximum speed. The excellent road-holding and unusually moderate thirst for fuel remain.

The original 444 had a 1,414 c.c., o.h.v. engine developing 51 b.h.p. Then, with the introduction of the Amazon saloon, came the 1.6-litre unit which had a higher power output. For the Volvo sports car the compression ratio was raised and twin S.U. carburettors were used. Now this engine, with a claimed power output of no less than 85 b.h.p., is fitted to the 444 primarily for export. In a car weighing 19 cwt ready for the road with five gallons of fuel, a good performance can be expected; with two people and the road test equipment aboard the b.h.p. per ton figure worked out at 75.7.

The shortcomings resulting from use of this more highly tuned engine are few. The slow running is a little rougher, warming up is rather slow even when the optionally extra radiator blind is used, and three or four times during the test (no more) the carburettors coughed a little when the engine was warm. The impression was formed that the carburettor needles were not *quite* right yet, and that the mixture was, therefore, very slightly weak in certain conditions.

However, the performance proved to be exciting in a family saloon of this size. At a mean 20.1sec the Volvo nearly "beat the 20" on the standing quarter mile (it achieved that in the favourable direction). It reached 30 m.p.h. in 4.6sec, 60 in 15.9sec and 80 m.p.h. in a mean of 35sec. For a car of this type to achieve a genuine 80 m.p.h. at all is creditable, but the Volvo went on quickly to a substantially higher speed, and exceeded 90 m.p.h. in one direction.

There are not very many saloon cars, even of much larger engine size, which can keep pace with the Volvo on

Below: Twin S.U. carburettors distinguish the latest 85 b.h.p. engine. Attached to the bulkhead is the fuel consumption meter used for The Autocar's tests. Right: All instruments and minor controls are directly in front of the driver. They include radiator blind adjustment, water thermometer, oil pressure and fuel level gauges and an ammeter. Controls for the heater are on the left.

European roads. On one journey of very mixed motoring conditions, taking in winding roads in the Hertford area, 46.7 miles were put into the hour, and this with a m.p.g. figure of 28.2. This suggests, rightly, to those who know the district, that the suspension, steering and brakes are of a very high standard—much better than those of the great majority of family models. One of the best features of the car is the way in which any harshness in the springing has been avoided while stability and freedom from roll remain impressive.

On the debit side are the slightly dated body shape, limited rearward vision through a window that is very shallow (even though the range of view is satisfactory), and seats which, while giving very good location for driver and front passenger, are a little hard. Compensating factors are that this coachwork is known to be long lasting, that it is clearly put together carefully and that it would be easy to repair in the event of accidental damage.

In conjunction with the power output, the choice of all ratios in the three-speed, central change box is ideal. While its flexibility on the overrun in town seems somewhat below average, the car will get away quickly in second from a crawl and reach a maximum of just over 60 m.p.h. in the same delightful ratio. The pulling power on hills in any gear is admirable.

The new engine, coupled with the already known and respected handling qualities, makes this export Volvo especially well suited to the man who would like best to run a sports car, but must have a four-seater saloon for family reasons. While the car is easy for anyone to drive, the skilled motorist gets, in addition to high performance, great satisfaction for the way in which that prowess is achieved.

VOLVO PV444 WITH SPORTS ENGINE

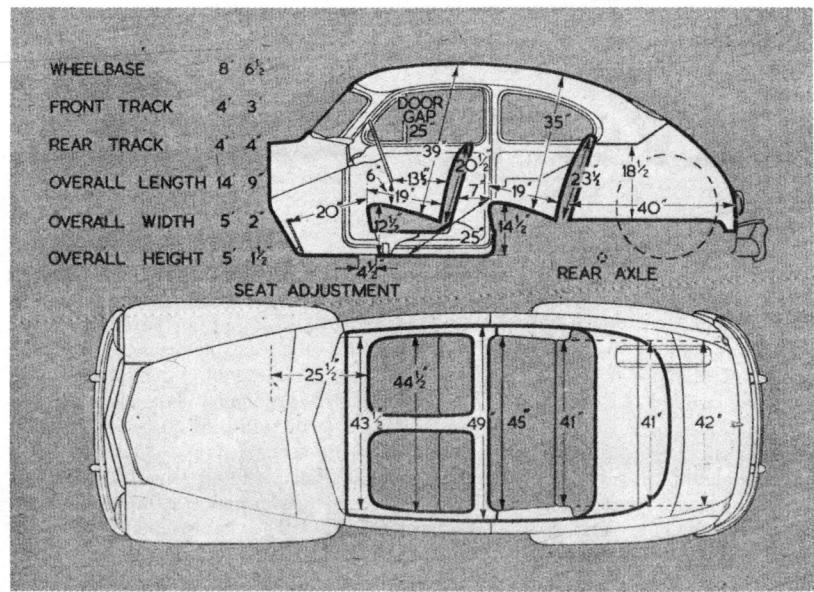

Measurements in these ⅛in to 1ft scale body diagrams are taken with the driving seat in the central position of fore and aft adjustment and with the seat cushions uncompressed

PERFORMANCE

ACCELERATION: from constant speeds
Speed Range, Gear Ratios and Time in sec.

M.P.H.	4.55 to 1	7.40 to 1	14.25 to 1
10—30	—	6.2	3.7
20—40	10.5	5.9	—
30—50	11.1	6.4	—
40—60	11.9	7.7	—
50—70	13.8	—	—

From rest through gears to:

M.P.H.	sec.
30	4.6
50	11.2
60	15.9
70	23.5
80	35.0

Standing quarter mile, 20.1 sec.

SPEEDS ON GEARS:

Gear	M.P.H. (normal and max.)	K.P.H. (normal and max.)
Top (mean)	90	144.8
(best)	91	146.4
3rd	47—61	75.6—98.2
2nd	24—32	38.6—51.5

TRACTIVE RESISTANCE: 20 lb per ton at 10 M.P.H.

TRACTIVE EFFORT:

	Pull (lb per ton)	Equivalent Gradient
Top	215	1 in 10.4
Second	381	1 in 5.8

BRAKES:

Efficiency	Pedal Pressure(lb)
25 per cent	25
45 per cent	50
90 per cent	75

FUEL CONSUMPTION:
32.8 m.p.g. overall for 327 miles. (18.5 litres per 100 km.)
Approximate normal range 28.2–43.8 m.p.g. (10.1–6.4 litres per 100km.).
Fuel, premium grade.

WEATHER:
Air temperature 85 deg F.
Acceleration figures are the means of several runs in opposite directions.
Tractive effort and resistance obtained by Tapley meter.

SPEEDOMETER CORRECTION: M.P.H.

Car speedometer:	10	20	30	40	50	60	70	80	90
True speed:	10	18	27.5	37	46	56	66	75	85

DATA

PRICES: To be announced.

ENGINE: Capacity: 1,580 c.c. (96.4 cu in).
Number of cylinders: 4.
Bore and stroke: 79.37 × 80mm (3.13 × 3.15in).
Valve gear: o.h.v., pushrods.
Compression ratio: 8.2 to 1.
B.H.P.: 85 at 5,500 r.p.m. (B.H.P. per ton laden 75.7).
Torque: 86.8 lb ft at 3,500 r.p.m.
M.P.H. per 1,000 r.p.m. on top gear, 16.2.

WEIGHT: (with 5 gals fuel), 19¾ cwt (2,177 lb).
Weight distribution (per cent): F, 52.7; R, 47.3.
Laden as tested: 22½ cwt (2,513lb).
Lb per c.c. (laden): 1.5.

BRAKES: Type: Wagner.
Method of operation: hydraulic.
Drum dimensions: F, 9in diameter; 2in wide. R, 9in diameter; 2in wide.
Lining area: F, 72 sq in. R, 72 sq in. (128.3 sq in per ton laden).

TYRES: 5.90—15in, tubeless.
Pressures (lb per sq in): F, 21; R, 24 (normal).

TANK CAPACITY: 7¼ Imperial gallons.
Oil sump, 6¼ pints.
Cooling system, 17 pints.

TURNING CIRCLE: L, 33ft 6in. R, 34ft.
Steering wheel turns (lock to lock): 2¾.

DIMENSIONS: Wheelbase: 8ft 6in.
Track: F, 4ft 3in; R, 4ft 3¼in.
Length (overall): 14ft 9in.
Height: 5ft 1½in.
Width: 5ft 2½in.
Ground clearance: 7½in.

ELECTRICAL SYSTEM: 6-volt; 85 ampere-hour battery.
Head lights: Double dip; 45–40 watt bulbs.

SUSPENSION: Front, Independent, coil springs. Rear, Coil springs, torque arm. Anti-roll bar position Front.

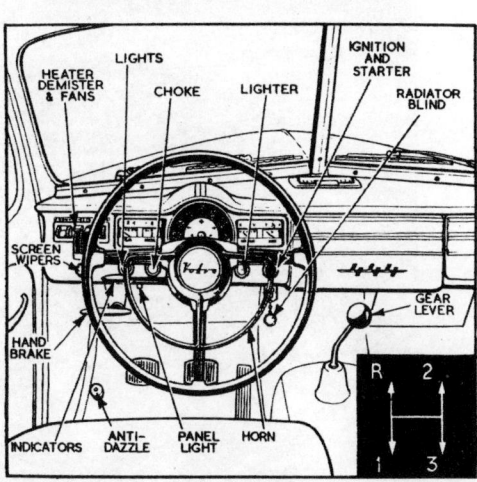

Often compared to a '41 Ford, the Volvo has a certain pleasing quality. Who knows, maybe the '41 Ford is coming back.

ROAD TEST VOLVO PV-444-L

ONLY A FEW MONTHS AGO, in April, we tested the 70-bhp Volvo. Now, along comes a real surprise, the same car with 85 bhp. Everyone remarks about the similarity of appearance between the Volvo and a 1941 Ford. Now we can add another Ford feature of that era, the 85 horsepower. Volvo called the 70-hp model the PV-444-K; the new 85-hp model is officially the PV-444-L.

The new engine (with a larger bore) is designated as the B-16-B, and already some sources are casting strong doubts as to the accuracy of the advertised bhp. Simply on the basis of a displacement increase from 1414 cc to 1577 cc, the power should go up from 70 to 78. But the compression ratio has been raised from 7.8 to 8.2:1, and this will add further to the output. Also, the torque peak now occurs at 3500 rpm (formerly 3000) which would indicate a camshaft change. Accordingly, we see no reason to doubt the ability of this engine to produce as claimed.

As a matter of fact, we essayed a rather extensive series of Tapley meter tests, toward the end of determining the exact rear-wheel horsepower. We were hampered by a low-speed carburetion fault, and the results were inconclusive. This much we do know: the Tapley readings of pulling power indicated more than the claimed 14.5% increase in torque.

The carburetion fault was corrected by our supplier (Ron Pearson, the invincible Volvo exponent) but even so, the cold figures show that the 0 to 30 and 0 to 40 times were not quite so good as before. This was hard to explain until we discovered that low gear has been altered slightly, from 3.23 to 3.13. The most impressive performance gain found is in high gear and above 60 mph. The improvement is shown graphically on the acceleration chart.

The average timed top speed proved to be 93.8 mph, or 3.8 mph more than in the earlier test. Such a speed is truly astonishing for a 1.6-liter sedan. A rough calculation shows that this increase in top speed would require 8 more bhp at the rear wheels. (Based on cw = .5 and A = 22 sq. ft.)

Putting all considerations of performance aside, the Volvo is still a tremendous automobile as a sturdy and practical

Chock full of machinery, the engine room shows no exterior change from the earlier 70-hp version.

Pleasing use of trim to compliment rather than to ornament the lines of the body . . .

Blue and cream plastic interior suggests a very expensive customizing job.

with 85 bhp, the sturdy Swede comes out swinging

utility sedan. When really thrashed the fuel consumption drops to 23 mpg, but normal 55/60-mph highway cruising will give 27 mpg as a best figure. It will cruise comfortably and easily at 75/80 mph, and under light throttle application the power unit is smooth and quiet. Unfortunately, the vigorous sports character of this unit becomes quite apparent when it is pushed hard. Under full throttle it seems to vibrate and becomes noticeably rough and noisy. With fond recollections of the 1931 PA Plymouth's smoothness, we fail to see why a small 4 should be quite so harsh as this one. Yet there is no question but that this is as tough a little engine as you will find anywhere, today.

Chassis-wise, the new Volvo continues with its proven unit construction. Road-rumble has been well subdued. As a matter of fact, the Volvo is not a light car (this one had a radio and weighed 50 lb more than our 70-bhp test car) and it uses heavier than normal gauge steel in many body and structural parts. The solid rear axle is located by a long rubber-insulated trailing arm on each side and uses coil springs. This and an equally well insulated front suspension of conventional design are responsible for an excellent ride, moderate roll, and generally good handling qualities.

The steering, as before, requires 3.2 turns and is light in action, with moderate understeer. Cornered really hard, there is perhaps more roll and a shade more caster return than a sports car driver would like, but a family car man (or woman) will never complain about this. At over 80 mph the steering seems to get "light" and is a little vague, but not so sensitive as to be frightening. Freedom from road shock transmission to the steering wheel is excellent.

Clutch action is unobtrusive, with no sign of slip at any time. The brakes were used fairly hard on several occasions. They, too, are eminently satisfactory. The 85-hp car has more brake lining area than the other version.

Externally, the new model can be identified by the tubular bumper guards at both ends and a new trim around the grille. The interiors are substantially unchanged, except that two-tone plastic upholstery tends to brighten things up considerably. A heater and defroster are standard equipment, but there was no opportunity to try these.

We understand that plans for producing the sports roadster and a five-speed gearbox have been completely abandoned, but with a car like this—who needs a sports car?

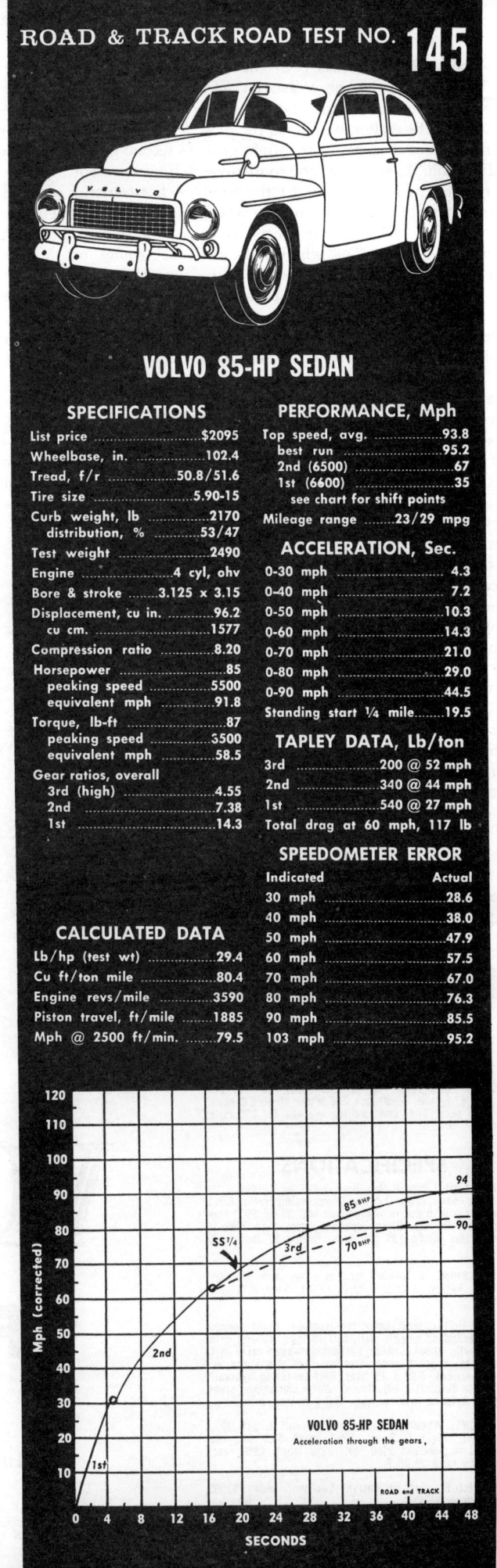

ROAD & TRACK ROAD TEST NO. 145

VOLVO 85-HP SEDAN

SPECIFICATIONS
List price	$2095
Wheelbase, in.	102.4
Tread, f/r	50.8/51.6
Tire size	5.90-15
Curb weight, lb	2170
distribution, %	53/47
Test weight	2490
Engine	4 cyl, ohv
Bore & stroke	3.125 x 3.15
Displacement, cu in.	96.2
cu cm.	1577
Compression ratio	8.20
Horsepower	85
peaking speed	5500
equivalent mph	91.8
Torque, lb-ft	87
peaking speed	3500
equivalent mph	58.5
Gear ratios, overall	
3rd (high)	4.55
2nd	7.38
1st	14.3

CALCULATED DATA
Lb/hp (test wt)	29.4
Cu ft/ton mile	80.4
Engine revs/mile	3590
Piston travel, ft/mile	1885
Mph @ 2500 ft/min.	79.5

PERFORMANCE, Mph
Top speed, avg.	93.8
best run	95.2
2nd (6500)	67
1st (6600)	35
see chart for shift points	
Mileage range	23/29 mpg

ACCELERATION, Sec.
0-30 mph	4.3
0-40 mph	7.2
0-50 mph	10.3
0-60 mph	14.3
0-70 mph	21.0
0-80 mph	29.0
0-90 mph	44.5
Standing start ¼ mile	19.5

TAPLEY DATA, Lb/ton
3rd	200 @ 52 mph
2nd	340 @ 44 mph
1st	540 @ 27 mph
Total drag at 60 mph, 117 lb	

SPEEDOMETER ERROR
Indicated	Actual
30 mph	28.6
40 mph	38.0
50 mph	47.9
60 mph	57.5
70 mph	67.0
80 mph	76.3
90 mph	85.5
103 mph	95.2

VOLVO 85-HP SEDAN
Acceleration through the gears

A PINCH OF SPICE has been added to the 1957 version of the Volvo PV-444. Outwardly the car retains the same appearance as the earlier model and the most prominent clue to a change is a slight grille modification. As we were soon to discover, the new bit of pepper is under the hood.

As you seat yourself behind the wheel, you will find that the car seems to be tailored to fit you. The wheel falls into your hands and you see over the top of the rim. Your feet readily find the pedals and the short-throw, floor-mounted gearshift represents the welcome return of an old favorite to many. There is no searching for information from the instrument panel. The instruments are grouped to be seen by the driver and the dial contrast is excellent.

It would perhaps be nicer if the front seats were so designed that the padding did not produce a rounded effect which gives a feeling of sitting *on* them rather than *in* them. In the back seat we found a roominess which pays tribute to the good space utilization in a compact design. The front seats are cantilevered back from a front mounting which leaves ample toe room for rear seat passengers. The wide doors are a great assist to getting in and out of all seats.

Getting back to the spice—the little four-banger under the hood has been increased from 70 to 85 bhp by virtue of a greater bore and increased compression ratio. You get an inkling of things to come when you turn the key starter and first hear the crisp, sharp and non-American exhaust note. A little patience is required after you start the engine because it has a slightly longer than average warm-up time.

As we wound the engine up in first and second gears we became aware of a marked improvement in the low-speed torque characteristics of the new engine. It is still necessary to keep the revs up to take advantage of what is in the engine; however, it "comes on" quicker. We proved this point on the drag strip. Our standing start to 60 mph time averaged 16.3 seconds which is 2.5 seconds better than we did with the smaller engine. To cover a quarter mile from a standing start took 20.3 seconds and we were doing 67.6 mph at the finish line. This was 4.6 mph faster than our speed last year.

Nothing has been lost of the excellent Volvo riding and handling characteristics in this new model. The ride is firm enough to give a feeling of stability without jolting hardness or wallowing mushiness. You come out of sharp dips at high speeds without bottoming or dangerous pitching. There is no excessive body lean in sharp corners and an absence of nose dive during very hard braking. These factors can be ascribed to the excellent snubbing and control of the coil spring suspension and the good weight distribution which holds the proportion on the front wheels to 52 per cent.

Good weight distribution is also reflected in the steering. Response to the wheel is quick and positive and there is neither a hair-trigger lightness nor a heavy feel at any speed. The three turns of the wheel lock-to-lock make the car most maneuverable and very easy to park. There is no hint of wheel fight, practically no road shock and if you release your grip on the

ROAD TEST

PERFORMANCE

Max. speed in gears, 1st 32 mph, 2nd 62 mph, 3rd 95 mph (factory rating). Acceleration: from standing start to 45 mph 10.1 secs., to 60 16.3 secs., ¼-mile 20.3 secs. and 67.6 mph, 30-50 mph 6.8 secs., 45-60 5.8 secs., 50-80 19.6 secs. Fuel consumption average for 280 miles 23.8 mpg.

SPECIFICATIONS

ENGINE: 4-cyl., in-line ohv. Bore 3.125 in. Stroke 3.150 in. Stroke/bore ratio 1.01:1. Compression ratio 8.2:1. Displacement 97.0 cu. in. Advertised bhp 85 @ 5500 rpm. Bhp per cu. in. .88. Piston speed @ max. bhp 2888 ft. per min. Max. bmep 135 psi. Max. torque 87 lbs.-ft. @ 3500 rpm.

TRANSMISSION: 3 forward speeds, 2nd and 3rd synchronized. Ratios: 3.13:1, 1.62:1, 1.00:1. Rear axle ratio 4.55:1.

CHASSIS: Unit chassis/body construction. Front suspension—independent wheels with coil springs, control arms and hydraulic shocks. Rear suspension—beam axle with coil springs, support arms, traction rods, track rod and hydraulic shocks. 5.90 x 15 tires. Self-centering hydraulic brakes with automatic adjustment. Worm and sector steering gear, 3 turns lock-to-lock, 13.9:1 overall ratio.

DIMENSIONS: Wheelbase 102.5 in., overall length 177.0 in., overall height 60.3 in., front tread 51.0 in., rear tread 51.8 in., weight 2180 lbs. (52% front, 48% rear), weight/bhp ratio 25.65:1.

PRICES (F.O.B. port of entry): Two-door sedan $2095, Station wagon $2345.

VOLVO
PV-444

An MT Research Report by Robert C. Scollay

COMPACT DESIGN places the car in class between large and small. Engine adjustment points and all service points are easily reached. Note the six-volt battery.

wheel in a turn, the car will return to a straight path.

Vision from the driver's seat can be considered good. Some slight interference is set up by the windshield corner posts, but on the other hand the flat glass in this V-type windshield does not have the usual corner distortion of a wraparound. The sweep of the electric wiper blades clears a most adequate area of the glass. The rear window is necessarily small because of the design, but here again it has only a gentle curvature and there is no evident distortion. Drivers will have to learn how to compensate for a small blind area at the rear right body corner.

To release the positive-type hood latching mechanism on the Volvo, you pull the handle on the firewall under the dash; reverse the operation when the hood is closed. Coupled with a front hinge, this should guarantee that the hood will never fly open while you are under-way. The engine compartment is neat and uncrowded. Minor engine adjustment points and the usual service points are reached very easily. We noted with interest that this is one of the few cars—especially among those of foreign manufacture—that retain a six-volt electrical system.

Luggage space in the trunk strikes a good balance with passenger space in the body. The spare is stowed at the extreme left and in an upright position. This leaves a good-sized flat floor area for other stowage.

Our general impression of the Volvo is that it is a good compromise between the small car and the so-called full-sized car. It is quick and agile in city traffic, and its good acceleration and top speed should keep you out of the laggard class on the open highway. Its size and weight are reflected in its gasoline consumption. We drove the car very hard and did everything in the book contrary to trying to obtain good mileage. We are aware that the tank average of 23.8 mpg could have been increased by at least five mpg had we been a little more conservative. In all departments, this is a good car.

DRIVING POSITION is comfortable, with the gear shift lever conveniently close to driver's right hand. Instrumentation is excellent.

SENSIBLE MOUNTING of spare contributes to usable space in the well proportioned trunk.

PHOTOS BY BOB D'OLIVO

Volvo's policy of constant product improvement is responsible for a beefier engine, a different grille, and stronger bumpers — all more in line with American taste. Interior of the car has been changed a bit, too.

SCI CAPSULE ROAD TEST:
A HOTTER VOLVO

THE FIRST Volvo to reach the U. S. was a subject of capsule road test in SCI for July '56. This car, driven by Ron Pearson, instantly dominated racing in Southern California in the under-1500 cc production sedan class. A year has passed, Volvo was beaten only once, and our old test car still is running like a watch. This is in spite of having been flogged in umpteen races and constantly thrashed as a sales demonstrator. It gives credence to the claim that 120,000 miles without a rebore is not unusual for these well-made Swedish cars.

Pearson's machine has been followed by over 11,000 Volvos on the west coast alone. They are not all the same by any means, because of the factory policy of steady improvement of the product, as opposed to design that is frozen for one or more model years.

The latest development in Volvo evolution is the retirement of the 86.7 cu. in., 70-bhp engine in favor of a 96.6 cu. in., 85-bhp version called the B16B. The increased displacement comes from enlarging the cylinder bore from 2.96 to 3.125 ins. The 21.4 per cent increase in power output results from the added inches plus an 8.2 compression ratio in place of 7.8 and two 1.5-in. S.U. carbs in place of a pair with 1.25-in. throats. A different camshaft evidently has been fitted, resulting in a less-flat torque curve that peaks with 87 lbs.-ft. at 3500 rpm rather than 76 at 3000.

What does the increased potency at the flywheel mean to the Volvo's performance? There is no difference below 40 mph but in the upper rpm range the transformation is radical. The zero to 80 mph time is reduced by 36 per cent and the car now surges up to 90 mph in almost ten seconds less time than the 70-bhp model takes to reach 80. Torque and pulling power are excellent in the new model's high speed range. In spite of the increased urge the difference in fuel consumption between the two models is insignificant. Engine noise and vibration are a bit more prominent in the 85-bhp model and it has a more pronounced tendency to run on for a few revs when the ignition is switched off.

The B16B engine develops its peak power at 5500 rpm but can be run up to at least 6400 without valve float. Our

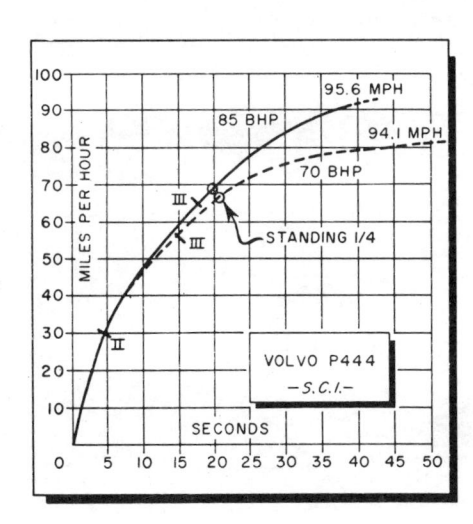

The B16B engine is rated 85 bhp. The higher output comes from a larger bore, a higher compression ratio, and bigger carbs. A new cam produces a steeper torque curve with a higher peak.

top speed of over 95 mph was clocked at 5800 rpm at the end of a 1.25-mile approach but revs still were mounting slowly. At top speed the car's handling is above criticism but from about 50 mph wind drums loudly in the passenger space if a window is open.

The Volvo recently has been given a beefier low gear but to engage it silently at a standstill it must first be "synchronized" by engaging top gear. The bumpers have been greatly improved and husky over-rider bars are standard equipment. The grille has been changed, heavy-duty electric windshield wipers have been adopted, and many detail improvements have been made to the car's interior. Points praised in last September's full-scale test report that deserve to be emphasized again are this machine's cornering ability and its welded body-frame structure, which feels as strong and solid as a steel safe.

The Volvo with B16B engine carries a port of entry base price of $2295 and represents an exceptional blend of low price, high performance, maneuverability, economy of operation, large carrying capacity, and low depreciation.

VOLVO PV444, 85 BHP MODEL

TEST CONDITIONS:
Number aboard 1
Temperature .... 73°F.

PERFORMANCE

TOP SPEED:
	85 BHP	70 BHP
Two-way average	95.0	94.1
Fastest one-way run	95.6	94.8

ACCELERATION:
From Zero to
30 mph .... 4.5 — 4.9
40 mph .... 7.7 — 7.5
50 mph .... 10.9 — 11.6
60 mph .... 15.2 — 17.3
70 mph .... 20.4 — 22.9
80 mph .... 26.5 — 47.2
90 mph 37.9
Standing ¼ mile .... 20.1 — 21.2
Speed at end of quarter .. 73 — 68

SPEED RANGES IN GEARS:
I .... Zero to 34 mph
II .... 10 to 65 mph
III .... 18 to top

SPEEDOMETER CORRECTION:
Indicated	Actual
30	28
40	38
50	48
60	58
70	68
80	76
90	86
100	94

FUEL CONSUMPTION:
Hard driving during accel. &
 speed runs .... 22.7 mpg
Average driving (under 60 mph) . 28 mpg

BRAKING EFFICIENCY:
1st stop 70
2nd 70
3rd 70
4th 70
5th 68
6th 67
7th 62
8th 55
9th 47
10th 46

POWER UNIT:
Type .... In-line four (three main bearing shaft).
Valve Arrangement .... Pushrod ohv.
Bore & Stroke .... 3.125 x 3.15 ins. — 79.4 x 80mm
Stroke/Bore Ratio 1.01 1.01/1
Displacement .... 96.6 cu. ins. — 1584 cc
Compression Ratio 8.2/1
Carburetion by .... Dual 1.5-in. SU side-drafts.
Max. bhp @ rpm .... 85 @ 5500
Max. Torque lb-ft @ rpm .... 87 @ 3500
Idle Speed 450

DRIVE TRAIN:
Transmission ratios
 I 3.13
 II 1.62
 III 1.00
Final drive ratio (test car) .. 4.55
Axle torque taken by .... Torque arms.

CHASSIS:
Wheelbase .... 102.5 ins.
Front Tread .... 51.0 ins.
Rear Tread .... 51.5 ins.
Suspension, front .... Independent, coil spring and wishbones. Anti roll bar.
Suspension, rear .... Coil springs, track rod.
Shock absorbers .... Double-acting telescopic.
Steering type .... Cam and two-stud lever (ZF).
Steering wheel turns L to L . 3.25
Turning diameter .... 33.5 ft.
Brake type .... Hydraulic; leading & trailing shoes F&R.
Brake lining area .... 116 sq. ins.
Tire size .... 590 x 15 (tubeless) Loaded radius 12.5 ins.

GENERAL:
Length .... 177 ins.
Width .... 62.5 ins.
Height .... 61.5 ins.
Weight, test car .... 2140 lbs (full fuel tank).
Weight distribution, F/R ... 52.4/47.6
Fuel capacity .... 9.5 U.S. Gallons

RATING FACTORS:
Bhp per cu. in. 0.88
Bhp per sq. in. piston area . 2.78
Torque (lb-ft) per cu. in. . 0.90
Pounds per bhp — test car .. 25.2
Piston speed @ 60 mph ...... 1910 fpm
Piston speed @ max bhp ..... 2890 fpm
Brake lining area per ton
 (test car) .... 108.5 sq. ins.
Mph per 1000 rpm 16.5

THE MOST amazing success story in the imported car field—which is a pretty amazing story itself—is the one that involves the Volvo. This is the car that makes no great claims to fashionable styling, but which has gone from scratch to the position of a sales leader in a very short time.

Right now, in fact, the Volvo is in second place among imported car registrations in California, the state that contains the largest share of the non-domestic car population. This also puts it ahead of some Detroit makes in local sales. Quite a feat.

There are, of course, several reasons for this sudden demonstration of popularity. One of them is the rather obvious advantage of being Swedish, since most anything coming from that country, including beauty contest entries, is highly regarded. Another is the rather brilliant handling of the Volvo by its distributors; the car's strong points have been fully exploited.

But the bulk of the answers, and those which are of first concern here, can be found upon examination of the car and what it has to offer. Recently, a few fundamental changes were made. They add up to marked improvement of existing characteristics, however, rather than any significant differences from what has been available in Volvo in the past. These changes can be pointed out during a general study of the car and the lineup.

THE LINEUP—The Volvo plant in Sweden turns out a wide variety of machinery, ranging from commercial and farm vehicles to both passenger and sports cars. Of all these products, only two are imported at this time. They are the two-door sedan and the station wagon.

The two-door sedan is identified by Volvo as the PV-444. It is larger than the average small imported car in what is generally called the economy class. But it should be kept in mind that despite this difference, the Volvo is an economy car and features excellent gas mileage, as well as low initial cost.

The utility vehicle is called the Volvo Duet station wagon. It also is much bigger than its continental contemporaries of this type. Other than the body, which strongly resembles in appearance the panel style of U.S. commercial vehicles, the wagon is basically the same as the PV-444 sedan.

(There have been more rumors about a sports car from Volvo than almost any other imported make of car. As yet, none have materialized in the U.S. in quantity. The factory admits that it is thinking in this direction, however, and it probably

VOLVO PV 444 sedan and station wagon are the only two models of the line now being imported to U.S. but have risen to second place in California foreign car registrations. Latest styling change is the mesh-type grille on the sedan above.

A Test—Guide

only is a matter of time before one appears in the states.)

PRICES—The sedan costs $2238 in New York at port of entry, the station wagon is $2490 including wheel trim rings, bumper guards, undercoating, directional signals, whitewall tires, heater and leatherette upholstery. Radio prices have to be determined locally. Prices on the West Coast are $2295 for the sedan and $2595 for the station wagon.

ENGINEERING—The body is the unit construction type. Engine is an ohv four-cylinder that has been boosted in the latest version from 86 cubic inches to 97 and from 70 to 85 hp. Compression ratio also is up, from 7.8 to 8.2-to-1. The gear ratio (transmission) for first gear has been lowered from 3.23 to 3.13, permitting more speed in low. The rear end gear ratio has not been changed. The rather fabulous second gear in the three-speed box remains the same; it can wind up to as much as 65 mph without protest. Dimensions of the car can be noted in the road test data box.

The design of the power train and running gear are quite conventional by U.S. standards. What sets it all apart and helps make the Volvo so attractive is the high degree of quality, precision manufacture and finish. It results in a vehicle that is exceptionally solid and sturdy, capable of very high performance and most reliable under rugged conditions.

STYLING—Not the latest look, obviously, but on the other hand there's nothing else like it on the road and this may actually help in some sales. At least there is an honest quality about its appearance. For the present version, there is a new grille, better upholstery and a rear side window that now swings open for ventilating.

VOLVO TEST DATA

Test Car: Volvo Deluxe two-door sedan
Basic Price: $2238 POE New York, $2295 West Coast
Engine: four-cylinder ohv 97.6 cubic inches
Compression Ratio: 8.2-to-1
Horsepower: 85
Torque: 87
Test Weight: 2315 lbs with driver
Transmission: three-speed manual
Cooling: water
Dimensions: overall length 177 inches, width 62.5, height 61.5, wheelbase 102.5, tread 51
Acceleration: 0-30 mph in 4.6 seconds, 0-45 mph in 9.6 and 0-60 mph in 14.2
Top speed: 90 to 93 mph
Gas mileage: 22.1 overall mpg

CARGO CAPACITY of Volvo wagon makes it suitable for either touring or delivery service with rear seat folded out of way.

INTERIOR and instrument panel are a bit austere by U.S. buyer's standards but the lively performance and economy sell the car once the prospect gets behind the wheel.

to VOLVO

FOUR CYLINDER ohv engine in the Volvo gets 85 bhp out of 97 cubic inches. Carburetors are British S.U.s. Top speed is over 90 mph and 0 to 60 is attained in 14.2 secs.

Many people have made the comment that "it looks like a '47 Ford that has been shortened and had a vertical section removed."

ROAD TEST—The test car was the two-door sedan, although drivers also covered quite a few miles in the station wagon to double-check results.

Most outstanding, naturally, are the performance figures. The engine refinements cut three seconds off the 0-60 mph time recorded in the Volvo of a year ago. Considering the number of cubic inches and the rated engine output, this is a fabulous accomplishment—and in a sedan yet! It takes a Detroit vehicle almost triple the displacement and horsepower to do very little better. Top speed is about 93 mph. And all this with gas mileage that will average 30 mpg with conservative driving. Is it any wonder that the car sells so well?

The Volvo rides firmly, handles very nicely. Some engineering improvements in suspension and rear axle supports help out. Maneuverability is excellent. Perhaps the strongest evidence in favor of the car's roadability is its record of road racing successes. In the sedan classes in the 1956-57 seasons, it rarely was defeated. Indeed, it was such success that brought it to the attention of the public and started the Volvo on its way to U.S. fame.

Drivers find all the conventional instruments on the dash. Seating in front is in bucket-type seats and very comfortable. Most novel, of course, is the floor-mounted gear lever—which incidentally can be downshifted at unusually high speeds without shedding parts all over the course. The shift pattern is standard.

Since the car is basically long and narrow, there is ample legroom. The greatest restriction is in three-abreast seating. Trunk space is most adequate. The seating arrangement of the station wagon resembles the U.S. system and can be noted in the accompanying photograph.

THE FUTURE—No major changes in the Volvo have been reported for 1958. Biggest changes are in the sales setup; there now are about 250 dealers in 36 states, plus the district of Columbia, Alaska and Hawaii. This is likely to increase. Service and parts rank with the best among all imports. Present owners are pleased and loyal.

Volvo seems to be building on solid ground. The car's popularity is expected to hold, although competition is getting more and more intense. But it is safe to say that when people discover the qualities of the Volvo, they like what they find and want it. •

How-to TOUR

You'd expect a small car to be a bit confining on the road. Yet adequate preparation—and a little care in planning the itinerary—can make even a long trip pleasant.

IN A TYKE

THREE thousand miles is a long way to go cooped within a space of some 30 cubic feet. Yet, this is about the room each of two passengers has inside the driving compartment of a small car, and that's what we had in our Volvo on a trip from New York to Sebring, Fla., and back. That's hardly enough area to stretch out in, much less live in for a steady thirteen hours. Yet we had only that much time at our disposal if we were going to make the start of the world-famous Sebring endurance run—and that's exactly what we meant to do.

However, as we discovered, certain preparations and plans before a trip can allay discomfort and subsequent fatigue. Preparations include: readying the car mechanically for the journey, careful packing, deciding approximate hourly timetable and route. Plans, on the other hand, involve the manner in which the distance will be traversed—that is, predetermining a method of driving.

After many miles and many mistakes, we learned many things about preparing and planning, and had we to do it over again, we'd proceed in the following fashion.

PREPARATION

Checking the vehicle:

Anxiety is one of the prime causes of fatigue. If the car isn't ready for the long hop, it may cause more than one apprehensive moment. An overhaul isn't necessary but these few provisions should be made:

The chassis should be thoroughly lubricated along with an oil and oil filter change. The latter will be of utmost importance to insure proper lubrication of the engine. Foreign matter in the oil can impede the flow with detrimental results. (We had such an experience on our trip, and we know.)

The fan belt should be inspected for frayed edges, tightened or replaced if worn—a generator won't charge without it. The output of the generator should be checked with the entire electrical system switched on. Headlight beams should be examined for brightness and adjusted to their proper level. At 60 *mph*, the distance of an upper beam is quickly overrun, even at correct setting. Kept too low it will prove eye-wearying and hazardous. Water hoses should be tightened and a minor tune-up performed. Tires should be inflated to recommended pressures or to two or three pounds above, since hard tires add to economy, stand up better under prolonged running.

So much for the mechanical end.

Charting the route

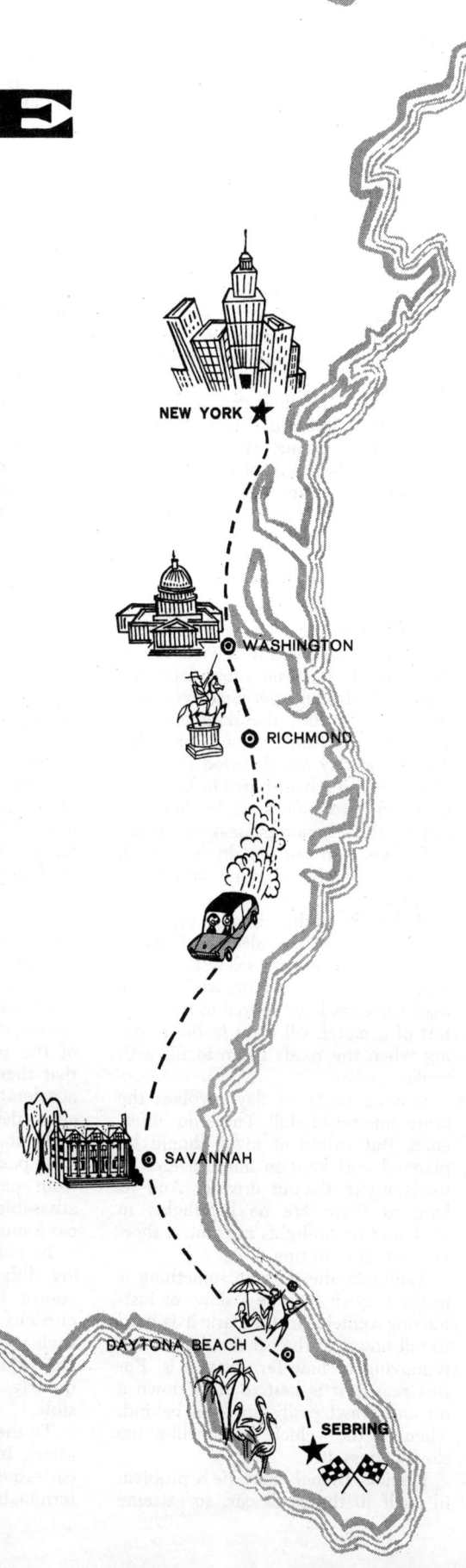

This preliminary is best left to the travel bureau of any oil company or the AAA. These agencies are advised of all detours, road constructions, blocks, congestions and will indicate them as such on the map. If any night travel is contemplated, it's well to know what's coming, where. On our trip we neglected this precaution and as a result lost time on a troublesome detour which could easily have been avoided.

Packing la petite

The amount of luggage taken is of course up to the individual. But regardless of quantity, the bulk of the baggage should be stuffed into the trunk of the car—not heaped upon the back seat. That area should be left free for loose clothes, pillow, thermos, and small necessities. The clothing is best spread out along the back seat or draped over the rear-seat backrest. Hanging suits and the like at the side-windows we found to be a mistake. They interfere considerably with side vision, cause squinting and neck-craning.

A large, soft pillow can be a blessing to the alternate driver. Propped on the seat against the backrest, it fits well in the small of the back and supports the entire torso—very much like a contour chair. Don't overlook this item!

PLANS

Driving for time

We found the best way to make time on the road was to drive at night. Done right, it is no more fatiguing than day driving—perhaps even less so. Traffic during the dark hours is almost non-existent. Towns crowded during the day are deserted at night. There are no school buses to halt traffic every two blocks while they discharge their minor passengers. There is far less stop and go driving which in itself is tedious, tiring and time-consuming.

However, night driving must be planned. It calls for sleeping through the day and traveling solely at night—from eight in the evening, when weary road travelers have retired to the comfort of a motel, till eight in the morning when the roads begin to fill with traffic.

Driving night or day involves the same amount of skill. There no difference. But riding at night should be planned and kept in mind. Deserted roads invite flat-out driving. And as long as there are no headlights in back and no taillights in front, a short fast run is tempting.

Taillights ahead mean something is in the way—it may be a slow- or fast-moving vehicle. In the dark it is hard to tell how fast the car or truck ahead is moving or how far away it is. For this reason it is best to slow down a bit and crawl easily up from behind. Then, if the vehicle is traveling too slowly, pass it.

Passing, though, may be a problem in itself in the small car, so extreme caution should be exercised. Before advancing, it is advisable to ride as close to the car in front as reflexes and brakes allow—with headlights on low beam. This reduces the gap between the vehicles, and the less the gap the quicker the pass. As soon as the left lane appears clear of approaching headlights, the car should be eased to the left, the upper beams flicked on to help illuminate more of the road ahead and to warn the overtaken car, and the throttle floor-boarded. This is all done on the assumption that the car has enough revs left to pass another fast-moving vehicle. For if the car is already at its upper limit, a passing maneuver would be ill-advised.

On the open road, exceeding the speed limit by a few miles might be considered unobjectionable. But hurtling through small, sleeping towns is inexcusable. And the local police usually lurk in dark places waiting to snare violators. The moral is, when in town, obey speed laws to the letter.

Driving for economy

Driving at night and driving for economy are practically synonymous. The less stop and go driving a car does, the more miles to the gallon it will deliver. Yet, all driving cannot be done after dark and since daylight driving and economy driving are antonymous, small pains must be taken if fuel is to be spared.

However, squeezing the most out of a gallon doesn't mean coasting down hills or handling the car like a sightseeing tourist. It simply means holding a steady speed, braking easily, accelerating slowly, and keeping the car within cruising limits. In towns, it means reducing the speed sufficiently to make as many lights as possible without having to stop. And when overtaking another vehicle, it means laying well behind so that a full view of the road ahead is visible, and so that there is ample room to work up acceleration. When the road looks clear depress the pedal slowly and make the overrun at an even rate. In this, passing by day radically differs from passing by night, when it is advisable to start the pass with your car is much closer to the vehicle ahead.

In another aspect, also, day driving differs. It is much preferable of course, if scenery, sights and side excursions are of interest. In our case, such matters were not on the agenda. Our aim was to reach Sebring as quickly and as economically as possible.

To the economy of our trip we can attest. In fact, by dint of inexcusable carlessness our gas ran low on an interminable stretch of road in the wilds of Georgia. We decreased speed from 70 to 30 and stayed on course. On our less than a gallon of gas we clocked a solid 26 miles before a gas station, open at that hour, finally hove into view. So more miles per gallon, characteristic of imported machinery, is not only a valuable economy feature. It can also save travelers like ourselves from their own inadvertency!

We made Sebring three hours ahead of schedule, which caused us to pass remarks to each other about how maybe we should have entered our little Volvo in the race, and also permitted us to catch a welcome nap before the start of procedings. On the return trip, thanks to the lessons acquired on the way down, we made even better time —and on three gallons less fuel!

What surprised us most of all—although probably it shouldn't have—was that the liveliness and roadability of our car eliminated much of the fatigue and feeling of monotony which would have occurred in a like trip in some Detroit giant. The handling qualities seemed to compensate for the comparative lack of room, and made the whole trip fun. Once again an imported machine demonstrated that above all else, and despite all rigors, it restores pleasure to driving—whether to the supermarket downtown, or to Sebring and back.

1. Touring in a small imported sedan does not have to qualify as an "adventure" if you make the proper preparations before you take off.

2. Small cars utilize all of the interior space, consequently they hold almost as much luggage as the false-fronted and finned native products.

3. While our Volvo didn't expend any oil during the 3,000 mile trip, it's a good idea to check oil level on small engines at every stop. High revving mills need proper lubrication to maintain efficiency.

4. Two Volvos are serviced down South. Simple four-cylinder engines used on most imports pose no problem for the majority of urban mechanics.

5. Individual bucket-type seats, and well-positioned steering wheel puts "you" back in the driver's seat. Long trip was fun in the roadable Volvo.

6. Passing finned monsters in a small sedan is an art that must be learned if you want to get the most out of your wee one during a long trip.

RAMBLER

By HRM Technical Editor RAY BROCK

Although this magazine is devoted largely to the fellow who understands automobiles and likes to tinker around with his car now and then to improve horsepower and performance. we are interested in the economy angle of an automobile, too. That's why we regularly print articles showing how to tune carburetion, ignition systems, etc., to provide maximum engine efficiency and produce the best mileage.

Letters from readers these past few months have revealed the fact that a lot of qualified hot rodders in this country are quite conscious of the increasing number of small economy cars on the road. We've had letters asking just how good some of these cars are and what shortcomings, if any, they might have. With an annual quiet period before the 1959 Detroit machines hit the showroom floors, we thought that this might be a good time to take a look at a couple of these small economy cars to see what they have to offer.

From the European scene we chose the Volvo, a car that not only fits in the economy category but also falls right in line as a performance model which does quite well in rallies and road races. From this country we chose the Rambler American which in our estimation is the only car currently being built in the U.S. that is in direct competition with the European imports. The reason for choosing these two particular cars is that they are very close in physical dimensions and construction, yet show a big difference between the European and American thinking on design and engine size.

Neither the Rambler American nor the Volvo are completely new automobiles for '58 nor for the past several years as far as that goes. The Volvo for '58 is practically the same as when the car was originally designed and built back in 1945. Only minor changes have been made to body, engine and suspension.

The Rambler first saw the light of day in 1950 and had only minor changes made until 1955. Through 1956 and '57, the small Rambler was not produced but for '58, the old bodies were dug out of the back room and dusted off with a few simple changes made to the sheet metal that brought the car up to date.

The Rambler American has a 100-inch wheelbase and the Volvo is 102½ inches in the same department. Overall car length from bumper to bumper is pretty close with the Rambler 178.3 inches long compared to the Volvo at 177 inches. Converted from inches to feet, both cars are just under 15 feet long. Width and height measurements show the more recent heritage of the Rambler since it is 57.3 inches high versus 61.5 for the Volvo and 73 inches wide compared to only 62 inches for the Volvo. Rambler wheel tread is 54.6 inches front and 55 inches rear, with the Volvo narrower at 51 inches front and 51.5 inches in the rear. Boiled down a little, the cars are nearly the same in length but the Rambler is a little shorter in height and a lot wider.

Both cars are built by the unitized construction method which means that there is no separate frame and body, rather a single welded unit with reinforced metal pads or short frame stubs where necessary to support engine or chassis components. This type of construction is very advantageous in the squeak and rattle department because it helps eliminate a lot of bolted or riveted joints between metal. The single unit construction is also quite strong and a lot of weight can be saved over the individual frame and body assembly method. The principal disadvantage to unit construction is in the body repair department in case of an accident but even this does not pose much

vs. VOLVO

Experienced in the small economy car movement, the Swedish Volvo and Rambler American take contrasting means to a common end.

Long favorite Rambler method of manifolding both intake and exhaust is simple, cheap. Intake passages are contained within head and block, exhaust pipe clamps directly to side of block.

Four cylinder Volvo engine uses twin carburetion, overhead valves and a fairly wild cam to get 85 horsepower out of the 97 cubic inch engine. Compression is 8.2; ignition, 6 volt.

Rambler interior is roomy, wide front seat comfortable for three. Instruments are all in one simple round cluster with switches, heater controls at the bottom edge of dash panel.

Volvo has elaborate instrument panel and floor shift for either three- or four-speed transmissions. Front seats are of bucket variety, fold well forward to permit rear seat accessibility.

Rambler American for '58 differs from its '55 predecessor with full wheel openings, more glass area, flatter roof, redesigned grille and a few chrome changes. Wheelbase is 100 inches and overall car length is just under 15 feet from bumper to bumper.

Rambler front suspension has coil spring hung between pad on inner wheel well of unit construction body and upper A-arm.

RAMBLER VS. VOLVO continued

of a problem today since body and fender men are used to welding body panels on the Detroit cars. The fact that nearly all U.S. cars will be of unitized construction in the near future is an indirect stamp of approval on the way both Volvo and American Motors have been building cars for the past several years.

SUSPENSION

VOLVO—If you were to put a Volvo on a grease rack and take a look at the chassis from the bottom side, you would be amazed at the similarity between it and some of the revolutionary suspension systems introduced on a number of our Detroit versions for 1958. The rear axle is a Spicer hypoid unit with the drive supplied by a two-piece open driveshaft. Rear suspension is by coil springs with stamped steel radius arms holding the axle in place. A cross-chassis anti-sway bar extends from the left side of the axle to the right side of the body. This much of the suspension is very similar in design to the '58 Lincoln and T-Bird and even partially reminiscent of Buick for the past couple of decades, minus the torque tube. Between the top of the differential housing and the underside of the body, a rubber bushed linkage is hooked in such a position as to eliminate axle roll due to torque and braking. This is similar to the '58 Chevy horseshoe shaped upper radius arm in location and function. The Volvo goes one step further than the Lincoln, T-Bird and Chevy however, with the addition of cotton webbing limit straps between the body and radius arms to limit the rear axle rebound over rough terrain. The Detroit method of using the shock absorber to limit the axle has been known to snap shock ends off.

The front suspension on the Volvo is similar in design to the American cars over the past several years. Unequal length control arms are used to maintain the proper wheel tread with a coil spring located between the lower control arm and a pad in the frame stub above. Steering linkage is ahead of the front wheels á la '58 Chevrolet. The front shock absorber is mounted between the upper control arm and the spindle support in an arrangement similar to that used for many years by Chrysler cars.

RAMBLER—The rear suspension on the little Rambler differs from that used on the larger models. Where the larger Ramblers use coil springs for all four wheels, the American rear suspension uses semi-elliptical leaf springs mounted in a conventional method parallel to the centerline of the car similar to most pre-'59 Detroit cars. The driveline is open and one-piece.

The front suspension is typical of that used by Nash and Rambler automobiles for years but unlike any other U.S. car. Unequal length control arms are used to maintain front wheel alignment and geometry but the coil spring used is mounted between a pad in the inside fender well of the unit construction body and the top of the spindle support above the upper control arm. This spring mounting places spring support directly in line with the wheel instead of between the lower control arm

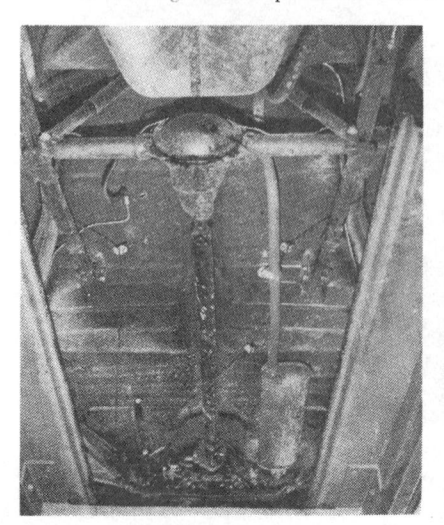

FAR LEFT. Rambler rear suspension is conventional but with front of semi-elliptical leaf springs bracketed to reinforced section in underside of unit body floor.

Volvo rear suspension resembles several current U.S. cars with coil springs, trailing radius arms, cross-chassis anti-sway bar and two-piece driveshaft to rear axle.

Photos by Eric Rickman

Volvo front suspension is conventional with coil spring between lower control arm and frame stub. Brakes have 9 in. drums.

Rear of Volvo is well streamlined for high speed driving. The wheelbase is 102½ inches and the overall length is 15 feet. Rear window angle and height cuts down visibility to the rear but other vision is good. Volvo is 62 inches wide and 61½ inches high.

and the frame. Shock absorbers are mounted between the lower control arm and the short frame extension from the unit body. Steering linkage, like that on the Volvo, is located in front of the control arms with the steering arms shaped to extend outboard of the backing plates to give the proper steering geometry.

ENGINE, TRANSMISSION

VOLVO—In keeping with the European trend toward small displacement engines, the Swedish engineers have come up with a four cylinder engine that has a 3.125 bore, a 3.150 stroke and a grand total of 97.6 cubic inches. The valve mechanism is overhead and actuated by rocker arms with a compression ratio of 8.2 to 1. Two SU carburetors are used to supply the fuel mixture and the engine is rated 85 horsepower at 5500 rpm with 87 foot/pounds of torque at 3500 rpm.

The car is available with only the standard floor shift transmission but is currently being produced in both the three-speed and four-speed versions. Both are listed as standard equipment with no extra charge for the four-speed but Volvo distributors do admit that the four-speed gear box will probably be found mostly in the deluxe models. Ratios for the three-speed transmission are 3.13 in first, 1.62 in second, direct in third and 2.66 in reverse. The four-speed gear ratios are 3.45 in first, 2.18 in second, 1.31 in third, direct in fourth and 3.55 in reverse. More about the advantages of each transmission a little later. The rear axle ratio with either transmission is 4.56 to 1.

RAMBLER—The Rambler has a little engine, too, but only when compared to U.S. standards. There are six cylinders in a row, each with a 3.125 inch bore and a 4.250 stroke for a total of 195.6 cubic inches. Just a little better than twice the number of cubic inches in the Volvo. The valve arrangement is in the block or you may prefer to use a phrase from out of the past, "It's a flathead." This engine was first introduced back in the 'forties and has been made in the 195 inch version since 1952 with only a couple of minor changes in compression. The present compression ratio is 8 to 1 and the horsepower rating is 90 at 3800 rpm with a torque reading of 150 foot/pounds at 1600 rpm. A single bore Carter carburetor takes care of the intake demands.

In the transmission department, the Rambler is like nearly all of the other cars built in this country. You can have your choice of overdrive or automatic at extra cost or the three-speed transmission. The standard three-speed gear box has ratios of 2.61 in first, 1.63 in second, direct in high and 3.54 in reverse. Overdrive ratio is .70 or a 30% reduction in engine rpm's at a given speed. The optional automatic transmission is titled the Flash-O-Matic and is made by Borg-Warner. It is three-speed with a combination of torque converter and planetary gears. The ratios are 2.40 in first, 1.47 in second, direct in third and 2.00 in reverse.

Rear axle ratios are also many and varied to match the transmission chosen or the owner's driving habits. For the standard transmissions, a 3.78 ratio is standard with 3.31 optional. With overdrive, 4.11 is standard, 3.78 optional. With the Flash-O-Matic transmission, the 3.31 gear is the only ratio available.

EASE OF DRIVING

VOLVO—They say that the first impression you get as you slip behind the wheel of an untried car is a lasting one and if that is so, then we have a complaint to lodge against the Volvo engineers. They put the steering wheel too far to the right—or maybe they put the seat too far to the left. Anyway, the wheel is about three inches to the right of the driver's belt buckle and when both hands are used on the wheel, the feeling that you are reaching off over thataway is disconcerting. A pair of bucket seats are used up front and with the car only 50 inches wide between the doors, plus a driveshaft tunnel between the seats, both driver and front seat passenger are pretty close to the door on one side. This closeness is not particularly noticeable until you start wheeling through some winding corners and find that the natural tendency to lean into the corner cannot be accomplished in one direction due to the door holding you back.

Shifting the transmission in the Volvo, whether three-speed or four is quite easy and sure with the clutch action very good. All four forward speeds of the four-speed transmission are synchro-mesh and downshifts are a snap when the need arises— which happens quite often with the low torque Volvo engine. Reverse gear in the four-speed unit requires a Houdini to locate it and a Boston strong boy to get it engaged. A husky detent prevents accidental engagement of reverse but almost proved too much for several of the office girls to manage. The

PERFORMANCE

VOLVO		RAMBLER
4.9 secs.	0-30 mph	4.3 secs.
15.8 secs.	0-60 mph	16.6 secs.
11.0 secs.	30-60 mph	11.9 secs.
20.3 secs.	¼ mile	20.5 secs.
70 mph est.	¼ mile speed	70 mph est.

MILEAGE

28.8 mpg	60 mph average	25 mpg
25 mpg	city driving	22.5 mpg
21.7 mpg	hard driving	21.9 mpg

three speed gear box is synchro-mesh only in the top two gears and shifting into first gear on a steep hill requires a skillful double-clutch artist or a complete stop to get the job done.

Only three turns of the wheel are required to turn from full right to full left and driver effort is at a minimum. The Volvo is very maneuverable and will cut a turning circle to either direction in just 35½ feet. The quick steering ratio is tricky for a driver fresh from a late model Detroit car but after a few miles behind the wheel, the action becomes more familiar.

RAMBLER—The first impression behind the wheel of the Rambler was that the car was not nearly as small as it looked. Steering wheel position was good and the front seat seemed nearly as wide as in a regular full sized sedan. Seating position was very comfortable.

The column shift lever for the standard transmission on the car we tried was too close to the wheel when low and reverse gears were being engaged but this should be easy to change. Rambler shift linkage was positive and clutch action smooth. Interior paneling on the Rambler firewall managed to get hooked by the toe of our number twelves the first few times we used the clutch but we discovered that using the ball of the foot instead of the arch eliminated that problem. Drivers with smaller feet probably wouldn't be bothered but the possibility is there.

Four turns of the steering wheel are required to turn from lock-to-lock in the Rambler. The steering ratio is quicker than that used on larger cars but not as quick as that on the Volvo. Effort needed to park the Rambler wouldn't put much strain on any size driver, male or female, and the car will turn around in a 36-foot diameter circle.

RIDE - HANDLING

VOLVO—Advertised as a family sports sedan and effectively proven in sports car races throughout the world, the Volvo has the semi-rigid suspension that accompanies a successful road race machine. Positive spring action and shock control do not combine to make this a boulevard limousine in the soft ride department. The car handles very well as it has proven in competition, but with the narrow wheel tread and high body, a sensation far different than that of the Detroit cars with excessive body lean was evident to us uninitiated. There seemed to be quite a bit of weight transfer in the corners and the rear end of the car doesn't appear to be in agreement with the front as to which direction they are going. Their record proves us to be too timid perhaps but, after all, the car was just borrowed and we didn't want to take it back in a basket.

For a sedan the Volvo is not too heavy, weighing in at 2150 pounds ready to run but minus passengers. On an open highway subject to gusty crosswinds, the Volvo is only average in stability despite its firm suspension. The light weight coupled with the high side view to catch the breeze probably account for the need to correct occasionally.

RAMBLER—This car is definitely not built for road racing and will protest when pushed through a corner at good speed by heeling over quite noticeably. It is no worse than the average American car however and can be crowded through some tight corners with fair speed should the occasion arise. In the ride department, the Rambler is a real smoothie. Unlike the Volvo with its firm suspension and choppy ride over small bumps, the Rambler cruises easily down highway or city streets with a minimum of discomfort to passengers.

There is more weight to the Rambler than the Volvo so this, too, contributes to the ride with a better sprung versus unsprung weight ratio. The Rambler will tip the scales at about 2595 pounds full of fuel but without passengers, almost 450 pounds over the Volvo. Stability in sidewinds was very good and a minimum of steering wheel corrections were needed.

PASSENGER COMFORT

VOLVO—With bucket seats, only two people can occupy the front seat of the car and since the car is advertised as a five passenger car, they must plan on three people in the back seat. Three of the fellows in the office who occupied the rear seat one day on the way to lunch will argue vehemently against the five passenger rating unless the rear seat occupants are in the younger age bracket. Rear seat width at the narrowest portion is just 42 inches but foot room is very good and head room is ample for a six footer. There is no rear seat ventilation on the cheap models but the deluxe version has the rear side windows hinged on the front edge and they swing out at the rear about three inches to improve circulation. Circulation around the driver's feet is restricted to a duct about ¼-inch wide and four inches long. Hot weather driving produced very hot feet.

Two types of upholstery are used, the regular cloth type and a vinyl type stain resistant covering. Our test car had the optional vinyl upholstery and interior workmanship was very good but we noticed that after driving for any length of time in hot weather, your pants and shirt wanted to stay in the car when you got out. The Vinyl upholstery does not permit any ventilation to the backside of the passengers and the seats become very uncomfortable on long drives in hot weather.

The Volvo is not by any means a quiet car to ride in. For some reason, the exhaust system has been designed to sound like a noise maker from Spike Jones' band and each time the throttle is released to change gears or decelerate, it sounds as though somebody behind you is giving you the raspberry. With the 4.56 rear axle ratio and a rocker arm engine up front, highway cruising speeds require lots of engine rpm's and the engine noise is quite noticeable to passengers.

Vision from the Volvo is quite good in all directions except to the rear. When using the rear view mirror, the small glass area plus the flat angle at which it is mounted high in the sloping body shell make it hard for the driver to tell what is following, especially if the trailing car is slightly off to one side.

RAMBLER—Also designated a five passenger sedan, the Rambler goes about it in a little different way by using the rear seat for two passengers and putting three up front where the seat is wider. The rear seat is 45¼ inches wide which is plenty for even a couple of middle-aged dowagers. The front seat is divided with two-thirds to the right side which folds forward to permit easy rear seat entry and exit. The front seat cushion is 58 inches, equal to most full sized American cars, and three adults can occupy the space quite comfortably. Head and foot room both in the front and rear seats is ample and in fact much better in the rear seat than in most of our full-sized hardtop coupe models.

The rear windows roll down about four inches for ventilation, but in the leg area up front the Rambler is very little better than the Volvo. There is no large fresh air duct for the feet on hot days but just the fresh air supply that makes its way through the heater ducting. The cloth upholstery used in the Rambler is both attractive and comfortable. We experienced no stickiness in 110° San Fernando Valley heat during a recent hot spell and although ice cream or candy might quickly leave stains, you can always install seat covers to hide the dirt later on.

The high rear axle ratio of the Rambler plus the quiet running L-head engine make engine noise unnoticeable when cruising down the highway at any speed. The absence of noise in the Rambler is equal to almost any car on the road and makes the car very pleasant to drive on long trips.

Vision in the Rambler is superb. When the American body dies were dug out of the moth balls for '58, one of the changes made was in the rear window department. The rear glass was expanded to a total of 700 square inches, just 40 short of the windshield, and makes all-around visibility in the Rambler comparable to an aircraft control tower.

PERFORMANCE

VOLVO—The small four cylinder engine that powers the Volvo is typical of the European economy engines. Small displacement but capable of a pretty fair horsepower per cubic inch ratio when twisted up to the higher rpm's. To get 85 horses out of only 97 cubic inches, the Swedes have really hopped the Volvo engine. The camshaft is undoubtedly a very radical grind and the car idles roughly, a matter which is accentuated by the spongy motor mounts used. Dual carburetion is also used to provide a better distribution of fuel at high rpm's. Peak horsepower at 5500 rpm and only 87 foot/pounds of torque at 3500 rpm show that this engine has to be kept screaming to produce power. With the four-speed gear box, engine rpm's don't drop too low when a shift is made and the engine will accelerate the car at a good clip. The gear ratio spread in the transmission is not the best with a long gap between the 2.18 second gear and the 1.31 third so the car will falter a bit when going from second to third.

Acceleration tests between the Volvo and the Rambler as well as checking them individually with the stopwatch showed that the cars were very close together in performance but each car went about getting there in an entirely different manner. The Volvo had rpm's and the Rambler had torque.

City driving in the Volvo revealed that a lower gear had to be used in many instances after slowing down to take a corner. Actually, the three-speed transmission with a 1.62 second gear ratio proved to be better for city driving than the four-speed box. In many cases, the 1.31 ratio of third gear was not low enough to provide the proper acceleration while the 1.62 ratio was. In residential sections with steep hills, however, the four-speed unit with full synchro-mesh gears was far easier to manipulate.

There is no need to baby the brakes on the Volvo when coming back down out of the steep hills because they are excellent. The pedal is very firm and requires a more than average amount of pedal pressure but the brakes stop true time after time without fading. Even when the brakes were so hot that the brake linings were smoking, stops were made without excessive pedal pressure or swerving. Nine inch brake drums are used on all four wheels with 144 square inches of lining area.

RAMBLER—As we mentioned earlier in the story, the Rambler engine is small by U.S. standards. With only 195 cubic inches, it moves the car around quite well. The horsepower rating is 90, just five more than that of the Volvo but notice where the maximum horsepower and torque readings were made. The Volvo peaked out at 5500 rpm while the Rambler six reached its maximum at 3800 rpm. As for torque, the Volvo had 87 at 3500 rpm while the Rambler gets almost twice as much, 150 foot/pounds, at less than half the engine speed, 1600 rpm. In other words, the Rambler engine is a "lugger" and practically eliminates the need for shifting gears because it will almost take off from a dead stop in high gear. With all the low speed torque and the single small carburetor, though, the Rambler engine naturally falls flat on its face if you try to twist it up to high engine speeds where the Volvo starts to feel healthy. There is absolutely no need to use second gear for acceleration after rounding a tight corner, the long stroke six cylinder engine pulls away quite nicely.

The brakes on the Rambler are nearly equal to the Volvo in size and lining area with nine inch drums used and 139½ square inches of lining area. Rambler brakes are also equal to the Volvo in stopping ability. Required pedal pressure is much less with the Rambler than the Volvo and brake sensitivity is nearly perfect. Rambler brakes are superior to most larger American cars that we have tested and would stop the car time after time without fading or pulling erratically. Even after repeated abuse, straight line stops were easy although more pedal pressure was needed and more stopping distance required.

ECONOMY

VOLVO—For a car that can cruise down the road at 75 mph without straining too much and actually be capable of an honest 90 mph, the Volvo is able to give some respectable mileage, at least in comparison to what we Americans are used to lately. Steady 60 mph cruising netted 28.8 miles per gallon on trips and mileage for city driving around sprawling Los Angeles was just about 25 mpg flat. The lowest mileage figures we got with the Volvo was 21.7 mpg for one tank of gas when we were doing acceleration tests and wheeling through one of the local twisting mountain regions.

The gas gauge in the Volvo has a habit of dropping from full to empty in a big hurry when on a trip but this is for the simple reason that the Volvo only has a 9¼ gallon capacity. Even at 29 mpg on a trip, the gauge creeps close to the empty mark after about 200 miles of running and you must stop to refuel. A larger tank would be a handy item, especially for trips.

Oil consumption during the 1000 miles we logged on the Volvo was nil. The twin carburetion seemed to be the only tricky point and the Volvo engine would idle alternately smooth and rough for no apparent reason.

RAMBLER—Steady 60 mile per hour highway driving in the Rambler netted a mileage figure of 25 mpg. City driving did not seem to seriously hurt the Rambler and average fuel consumption in the city was 22.5 mpg. Acceleration tests and mountain driving produced only slightly less mileage with a 21.9 mpg figure registered. We doubt whether any type of everyday driving would give less than 22 miles per gallon. For people who put lots of miles on a car out on the open road, the overdrive would probably add a couple of more miles to the gallon. The automatic would naturally rob a little gas around town but on a cross country cruise with the 3.31 axle ratio, even here the mileage should be fairly good.

Remember when the Rambler used to be called the Rambler Cross Country? Well, they haven't been using the name since they revived the 100 inch wheelbase model but the ammunition is still there. The name came from the fact that a twenty gallon fuel tank was used which coupled with the Rambler mileage potential permitted Rambler owners to take an average cross country trip without stopping to add gas. At 25 miles per gallon, the 20 gallon tank gives a range of 500 miles and that's as far as most people will drive in a day.

CONCLUSIONS

Driving the two cars we tested over a period of two weeks, alternating from first one to the other, we came up with some definite ideas about them. They are two wholly different types of cars although they are both about the same in exterior dimensions. The Volvo has a little engine that you have to keep buzzing through the use of gears to get performance. The high rpm's and rigid suspension plus a noisy exhaust and the fact that it is made in Europe qualifies the Volvo as a sports car and that is what the Volvo people call it. A trip through a few curves with the Volvo will have you thinking that you are Juan Fangio.

The Rambler American on the other hand has an engine that will turn just about half the rpm of the Volvo while doing the same amount of work. Gear shifting is not required as often as in the Volvo and the engine is quiet, making no claim of ability through exhaust noise. The Rambler is not capable of winning road races with its soft ride but won't tire passengers on a long trip. In the mileage department, the Volvo was almost 4 miles per gallon better than our test Rambler but overdrive could help cancel some of this advantage. Of course, an overdrive is $102 extra.

As for price, you would be surprised just how hard it is to get the real figures. Both cars are available in plain or fancy models and while you have to pay freight from Detroit on one, you have to pay freight from the port of entry on the other. If you are interested in either of these cars, we suggest that you check with your nearest dealer to get a price. They are both good cars although in entirely different respects and if you're tired of the 21 foot job from which you're getting 11 miles per gallon, take a look at the Volvo or Rambler American. Driving can be fun—and cheap, too.

ROAD TEST 4-SPEED VOLVO

THE VOLVO has steadily, since the date of its first appearance in the U.S., continued to make friends. In the past it has been because of some fine virtues: economy, performance, sturdy construction and rugged dependability, and in spite of some drawbacks: the resemblance to a 1940 Mercury with 1948 Ford fenders (all reduced in size), and a strictly American-style, 3-speed transmission. These latter "features," of course, may appeal to enthusiasts of older-model Fords and therefore would not be considered drawbacks by them.

Well, it still looks like the aforementioned hybrid refugee from Dearborn, but it now has a wonderful 4-speed, all-synchromesh transmission. Still operated by a long, floor-mounted lever, as in past models, it nevertheless is smooth in operation and adds to both the performance and the pleasure of driving the car. The only possible criticism of the transmission might be the spacing of gear ratios. Second and 3rd are just a mite too far apart.

This fault will not bother the driver in normal city driving, but only when maximum performance is desired or, occasionally, when ascending hills that may be too steep for 3rd yet not steep enough to require the use of the much lower 2nd gear.

When maximum speed in low gear is reached and a quick shift into 2nd is made, the car continues to accelerate in a brisk fashion. But when the car is all extended in 2nd gear and a shift made to 3rd, it bogs down a little due to the wide gap between the gear ratios. Then, with an upshift from 3rd to 4th, acceleration is steady once again.

In spite of the long floor shift lever, the gears were always easily engaged and the transmission is a real tribute to its designers. A remote shift in place of the long lever would be like having egg in your beer, and it should be possible without altering the passenger compartment at all, due to the individual front seats.

There has always been a certain amount of comment, from readers as well as the distributors of the cars tested, about the figures obtained during our road tests.

The tests conducted by Road & Track's staff are road tests, not destruction tests. Each car tested is driven as though it belonged to us, and therefore it is neither thrashed nor coddled. It is given as much consideration as our own personal cars. Accordingly, acceleration and gas mileage figures obtained by us should also be obtainable by an individual owning the same make and model car.

In the case of the Volvo, a mileage check resulted in an actual average of 25.8 miles per gallon in normal city driving. If the car were driven extremely hard this mileage

Engine and accessories are easy to get at and service.

Seat mounting leaves foot room for rear-seat passengers.

PHOTOGRAPHY: POOLE

Finer mesh grille is one of few changes in appearance.

a family-type sports car that really is

would fall off, and on the open highway, or if treated with extreme care, it would, of course, improve.

No changes have been made in the chassis/body of the Volvo, and apparently none are contemplated in the immediate future, so the handling and riding qualities have not changed from previous models (R&T tests, April 1957 and September 1957). Enthusiasts will welcome the added gear, though, as it does enable the driver to select a more suitable ratio to fit each occasion.

The engine still has the roughness that was mentioned in past tests (although to a lesser degree) but is one of the most free-revving rocker-arm engines we've seen. During the test, 80 mph (indicated) was reached in 3rd gear, and the car was still picking up speed when road conditions made it necessary to slow down. This works out to an actual 75 mph, at which the engine rpm was 6500, with no valve float indicated.

The company's claims for a family sports car are obviously not without justification, and if the prospective purchaser of an economy car is satisfied with the appearance of the Volvo he would be wise to give it consideration. There is ample reason to believe he will be happy with the car and can expect, and get, a long, trouble-free life from this Swedish product.

Bumper bracing is for U.S.-style bash and smash parking.

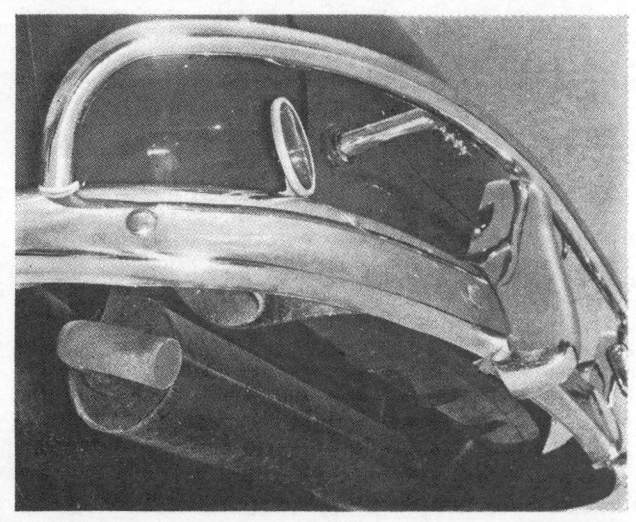

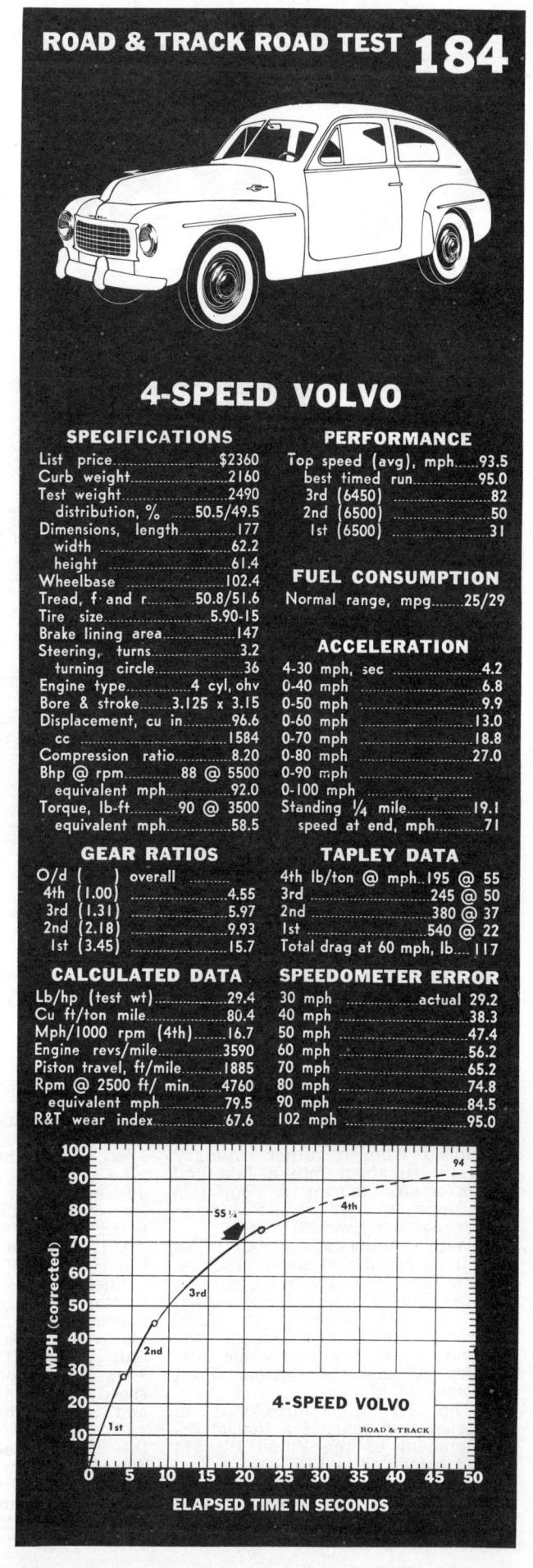

ROAD & TRACK ROAD TEST 184

4-SPEED VOLVO

SPECIFICATIONS
List price	$2360
Curb weight	2160
Test weight	2490
distribution, %	50.5/49.5
Dimensions, length	177
width	62.2
height	61.4
Wheelbase	102.4
Tread, f and r	50.8/51.6
Tire size	5.90-15
Brake lining area	147
Steering, turns	3.2
turning circle	36
Engine type	4 cyl, ohv
Bore & stroke	3.125 x 3.15
Displacement, cu in	96.6
cc	1584
Compression ratio	8.20
Bhp @ rpm	88 @ 5500
equivalent mph	92.0
Torque, lb-ft	90 @ 3500
equivalent mph	58.5

GEAR RATIOS
	O/d ( ) overall	
4th (1.00)		4.55
3rd (1.31)		5.97
2nd (2.18)		9.93
1st (3.45)		15.7

CALCULATED DATA
Lb/hp (test wt)	29.4
Cu ft/ton mile	80.4
Mph/1000 rpm (4th)	16.7
Engine revs/mile	3590
Piston travel, ft/mile	1885
Rpm @ 2500 ft/min	4760
equivalent mph	79.5
R&T wear index	67.6

PERFORMANCE
Top speed (avg), mph	93.5
best timed run	95.0
3rd (6450)	82
2nd (6500)	50
1st (6500)	31

FUEL CONSUMPTION
Normal range, mpg	25/29

ACCELERATION
4-30 mph, sec	4.2
0-40 mph	6.8
0-50 mph	9.9
0-60 mph	13.0
0-70 mph	18.8
0-80 mph	27.0
0-90 mph	
0-100 mph	
Standing 1/4 mile	19.1
speed at end, mph	71

TAPLEY DATA
4th lb/ton @ mph	195 @ 55
3rd	245 @ 50
2nd	380 @ 37
1st	540 @ 22
Total drag at 60 mph, lb	117

SPEEDOMETER ERROR
30 mph	actual 29.2
40 mph	38.3
50 mph	47.4
60 mph	56.2
70 mph	65.2
80 mph	74.8
90 mph	84.5
102 mph	95.0

Road Test

VOLVO: SWEDISH,

S.C.W. Scandinavian Correspondent von Tobiesen finds Volvo's latest a regular fireball. No wonder they sell in America...

Volvo designers are really safety conscious. Wheel is dished, dash and visors thickly padded. Front passengers have safety belts as standard.

by FRED VON TOBIESEN

AFTER the style of Cinderella, the Volvo PV 544 Sports developed from a rather ordinary family runabout into a fast and comfortable touring saloon, able to show its rear wheels to most cars on the Swedish roads.

With emphasis on safety as well as speed and sheer acceleration, the Volvo PV 544 Sports is a reasonable financial proposition in its home country at 12,250 Kroner tax paid. Cars in the same price range include the Riley 1.5 and the German Ford Taunus 17 M.

The PV 544, introduced in August last year, is a logical development of the Volvo PV 444 first presented in 1945.

A 40 b.h.p. o.h.v. engine powered the first Volvo; it was later developed to give 44 b.h.p. and eventually 51 b.h.p. A 60 b.h.p., o.h.v. four-cylinder unit of 1,580 c.c. replaced the previous 1,480 c.c. engine in 1957. This engine is still used and is now offered in two versions: the normal 60 b.h.p. unit, and the sports, which gives 85 b.h.p. at 5,500 r.p.m. It is the latter that powers the Sports PV 544.

The current PV 544 is produced in three series: a Standard, a Special and a Sports. The Special offers the choice of a three or a four speed all-synchromesh gear box. The Sports has the four speed box as standard equipment and a final drive ratio of 4.1 to 1.

The PV 544 under review had the 85 b.h.p. engine, fed by two SU H4 carburettors with simple pancake air filters. The compression ratio is up to 8.2 to 1 and there is a "hotter" camshaft. The combustion chambers are quite normal and the valve gear is conventional with pushrods and rockers. The generous power output is developed without fuss apart from a healthy intake roar to please the enthusiast.

Driven briskly, the PV 544 Sports is noisy, but driven gently as a normal family saloon the engine is not much noisier than a normally silenced unit. It is smooth at all speeds and running is sweet apart from some slight roughness at tick-over. Response to the accelerator is instantaneous and the cold starting device makes for easy starts on frosty winter mornings.

With the radiator blind in its closed position, the engine quickly reaches operating temperatures. Fuel consumption is modest.

The body of the PV 544 represents probably the last stage of

BUT NO ICEBERG

Personal transport for a millionaire? The Volvo would do anyone, for it even fits with the atmosphere of this lush Swedish hotel.

VOLVO: SWEDISH BUT NO ICEBERG

development of the basic Volvo design. Previously, criticism has been centred around the limited view from the interior, and the shallow rear window has been badly spoken of. Even after the latest increases in window areas —both windscreen and rear window—the PV 544 body cannot merit highest marks in this department. True, the windscreen has been sensibly enlarged and the old V-screen has been replaced by a one-piece affair with an excellent wiper arrangement. The pillars, however, are on the thick side and the side windows are somewhat small.

Seating is good. In front there are separate seats with the handbrake mounted on the transmission tunnel. At rear, a comfortable bench offers room for three at a pinch. The elbow rests are built into the body sides and there are two large ashtrays for the rear passengers. Safety belts are standard equipment for the front seats. In the back, fitting points for belts are provided. Belts are of the "across the chest" type.

A new facia covered with non reflecting p.v.c. is neatly finished and designed to provide some cushioning in the event of an accident. Similarly, the sun visors are of a shock absorbing material.

There is a reasonably large locker — lockable — and an ashtray as well as a lighter. Space is provided for a radio. The instruments are grouped in front of the driver. The speedometer is of the modern "thermometer" type and the coolant thermometer and the fuel gauge are grouped under the speedometer. Four warning lights give information about ignition, oil pressure, indicators and main beam. A trip recorder is standard on the Sports.

The pleasantly shaped steering wheel has a full circle horn ring with a lever controlling the indicators protruding from the steering column housing. If moved towards the steering wheel, the lever flashes the headlights. The electric wipers cover a large area and an electric windscreen washer is controlled by a second pull of the wiper switch. A comprehensive heating and demisting unit is also one of the standard fitments.

On the road, the well thought out seating is appreciated. It is possible to adjust backrest rake. The wheel is near vertical and the honest floor mounted gear lever gives excellent control of the four speed box. Legroom in front is plentiful and pedals are well placed. The accelerator is pendant.

The gear box with its four all-synchromesh speeds gives a special charm to the PV 544. Apart from a prominent whine in reverse, the box is silent. Second can be used for starting but bottom gear seems to be the natural choice. As full torque is delivered at 3,500 r.p.m., frequent use of the box brings out the full virtues of the power unit.

Top speed was not fully reached on public roads but an honest 95 m.p.h. seems to be possible and given favourable conditions, a genuine 100 m.p.h. would probably be recorded.

The brakes are smooth and perfectly even, inspiring confidence and giving strong and satisfactory retardation. For Swedish Winter

Rear view shows striking resemblance to early postwar Fords. Von Tobiesen thinks lamp units look clumsy, but likes improved rear vision

Volvo Sport has strangely archaic lines, yet it is attractive. Big windscreen is new for '59.

roads, they appeared to have that slight resistance to act too fiercely which is so important.

If brakes, clutch and gearbox were good, the headlights on the test car were not. Too short a distance ahead is covered by the lights to allow the performance of the car to be used at night. It it somewhat surprising, too, that the manufacturers have missed the opportunity of providing headlights with asymmetric dipping; this is a most useful European innovation intended to give more light along the near side with dipped lights.

Roadholding of the PV 544 is as near perfect as possible bearing in mind the dual qualities of the car. Damping is just about right and understeer is very slight. On gravel roads, the rear axle is perhaps a trifle too lively when the driver is enterprising but the PV 544 is easily corrected when the rear end does break away. At speed, the front end appears a shade too vague, but the car steers well even if a conscious effort is needed.

Rack and pinion steering would probably be better. Noises from road surfaces are well damped and wind noises at speed are eliminated if the rear windows are opened an inch or two and all fresh air is taken via the heater—hot or cold. The tyres are quiet and squeal is not normally provoked. #

Motor turns out 85 b.h.p. from only 1½ litres, yet the Volvo looks like any other family engine — apart from dashing paint job. Fully balanced SUs feed it.

SPECIFICATIONS

ENGINE:
Type, four cylinder in line, water cooled
Valve arrangement ...... pushrod o.h.v.
Bore and stroke ... 79.37 m.m. x 80 mm.
Cubic capacity .... 1,580 c.c.
Compression ratio .... 8.2 to 1
Carburetion by .... twin SU.
Max. b.h.p. at r.p.m.
        85 at 5,500 (gross)

CHASSIS:
Wheelbase .... 8 ft. 6⅜ in.
Front track .... 4 ft. 3⅝ in.

Rear track .... 4 ft. 3⅝ in.
Suspension front ... wishbones and coils
Suspension, rear ...... rigid, with coils
Steering wheel turns, L. to L. 3¼
Brakes .... Wagner hydraulic
Tyre size .... 5.90 x 15

GENERAL:
Length .... 14 ft. 5 in.
Width .... 5 ft. 3½ in.
Height .... 4 ft. 11½ in.
Ground clearance .... 7¼ in.
Weight as tested
   1,040 kgs. equals 20.4 cwt. approx.

PERFORMANCE

TOP SPEED (Estimated) 95-100 m.p.h.
SPEED IN GEARS:
I .... 30 m.p.h.
II .... 45 m.p.h.
III .... 78 m.p.h.
IV .... 95-100 m.p.h.

ACCELERATION:
From zero:
30 m.p.h. .... 4.0 sec.
50 m.p.h. .... 9.4 sec.
75 m.p.h. .... 24.1 sec.

(Full figures not given because of Swedish road conditions.)

Servicing the VOLVO

Larger SU Carburetors and More Horsepower Pose No New Maintenance Problem on B Engine

● Volvo, as is known to many owners, stands for "I roll" in Latin, and these Swedish 4-cylinder high-performance cars started "to roll" into this country in the spring of 1956. They "caught on" too, as within the space of a year they had copped seventh place in import sales—and their popularity is still growing by leaps and bounds.

Although Volvo has been part of the American scene since 1928, when it opened purchasing offices in Detroit, this industrial giant (one of the 100 largest non-U.S. firms in the world) did not tap the growing U.S. market for compact cars until 28 years later. Cars, of course, are not its only business, as it also manufactures trucks, buses, tractors, jet and marine engines, and printing presses.

The earliest model cars rolling into the U.S. and those imported up to about mid-1957 (8,035 in all) were equipped with 70-horsepower 86.6-cubic-inch engines. The subsequent and current models have a 96.4-cubic-inch (1580 cc) mill that develops 85 lively horses at 5500 rpm. The current model has a torque output of 87 lb. ft. at 3500 rpm, compared with 75.9 lb. ft. at 3000 rpm on the smaller engine.

In the middle of 1958, a 4-speed transmission with all forward speeds synchronized was made an available option. A hinged rear side window was also standardized at the same time. The biggest changes made in the 1959 PV 544 models were modifications including a larger one piece windshield with slimmer pillars, a wide and deeper rear window, a roomier rear seat and a new type speedometer and padded instrument panel. The handbrake was positioned between the front seats and new steering system, directional signals and accelerator pedal were also adopted.

Service accessibility is a strong point. The forward tilting hood, roomy engine compartment and straightforward design of the overhead-valve 4-cylinder engine contribute to the ease of servicing. The only underhood components that may look "strange" are the two British S.U. carburetors that contribute to the engine's high horsepower output and surprising performance characteristics.

The principal differences in the engines is that the early B 14 A type has a 2.953-inch cylinder bore and the later and current B 16 B powerplant has a 3.125-inch bore. Both engines have a 3.15-inch stroke; the smaller unit has a 7.8 to 1 compression ratio and the 85 horsepower model boasts an 8.2 to 1 ratio.

While the increased bore meant the use of different pistons and rings, the connecting rods are also beefed up. Readily seen, when comparing the en-

This is grille on older models. Note how it differs from that in photo on facing page.

Latest model (Volvo) incorporates a one-piece windshield with slimmer pillars, bigger rear window, wider rear seat, a four-speed transmission, brighter turn indicators, padded dash and new interior colors, fabrics.

gines, are a revised exhaust manifold and a relocated oil filter. However, other things include a different basic valve timing setting and valve tappet clearances, along with larger throat carburetors and a different ignition distributor and cooling system thermostat.

As the car has been in production so long it apparently is not subject to the problems that crop up in the frequently changed domestic models. Each engine is bench-tested at the factory before installation and then, when the car is complete, in a road test. Proper "break-in," as thoroughly outlined in the owner's manual, has an important bearing on engine life. While the basic running-in period is 900 miles, reasonable care should be exercised for the following 1250 miles.

Initial oil changes should be made at 500 and 1500 miles, and every 3000 miles thereafter. The oil pan should be thoroughly drained, flushed, cleaned and refilled at the 500 mile mark.

As the engine is a very responsive high-speed unit, drivers used to operating a domestic car with an automatic choke should train themselves not to overlook pushing in the choke when the engine is up to normal operating temperature. As with any car engine, the Volvo unit should not be raced when cold or when use of the choke is required for smooth running.

The 1956 and Blue 1957 Owner's Manuals on the 70-hp car and the Green 1957 Owner's Manual did not describe the operation of the British S.U. Carburetor. However, the 1958 and 1959 manuals have added 16 pages covering carburetor maintenance and setting of ignition. (This 1958 Form 656271 with valuable carburetor information is available from Paul Utans, Volvo Distributing, Inc., 452 Hudson Terrace, Englewood Cliffs, N.J. for $1.50 postpaid on a first-come-first-served basis until the supply is exhausted.)

The B 14A and B 16B engines are equipped with two coupled horizontal or side-draft carburetors that are unlike anything used on domestic cars. Other imports such as Jaguar, MG, Austin Healey and others use the same unit. The S.U.'s differ from domestic carburetors principally in the fact that instead of having a butterfly choke, accelerator pump and the usual variety of jets, they operate on a single variable jet device and an automatic expanding choke or venturi arrangement.

There is a conventional butterfly throttle on the engine side of the air throat, and the size of the throat is varied according to throttle opening, engine speed and load. The vacuum piston, besides controlling the size of the throat and the air flow, also meters the fuel by means of a tapered needle attached to its base. This fuel needle extends down into a jet sleeve (below the air throat) and regulates the fuel supply in proportion to the incoming air in unison with the movement of the piston and variation of air throat size.

VOLVO Continued

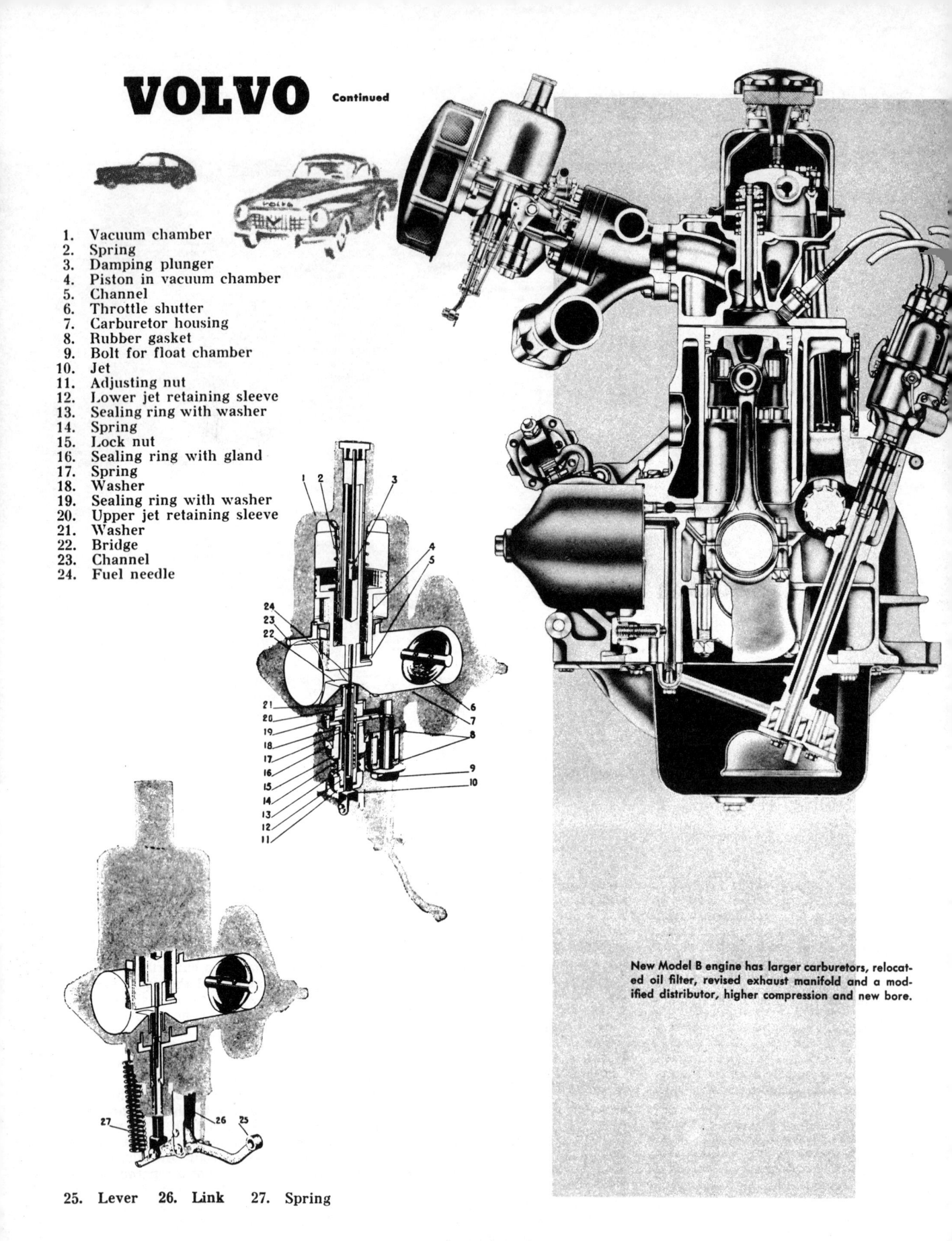

1. Vacuum chamber
2. Spring
3. Damping plunger
4. Piston in vacuum chamber
5. Channel
6. Throttle shutter
7. Carburetor housing
8. Rubber gasket
9. Bolt for float chamber
10. Jet
11. Adjusting nut
12. Lower jet retaining sleeve
13. Sealing ring with washer
14. Spring
15. Lock nut
16. Sealing ring with gland
17. Spring
18. Washer
19. Sealing ring with washer
20. Upper jet retaining sleeve
21. Washer
22. Bridge
23. Channel
24. Fuel needle

25. Lever 26. Link 27. Spring

New Model B engine has larger carburetors, relocated oil filter, revised exhaust manifold and a modified distributor, higher compression and new bore.

Volvos come equipped with twin SU carburetors which have only one jet, a variable jet controlled by a tapered needle responsive to engine vacuum.

62

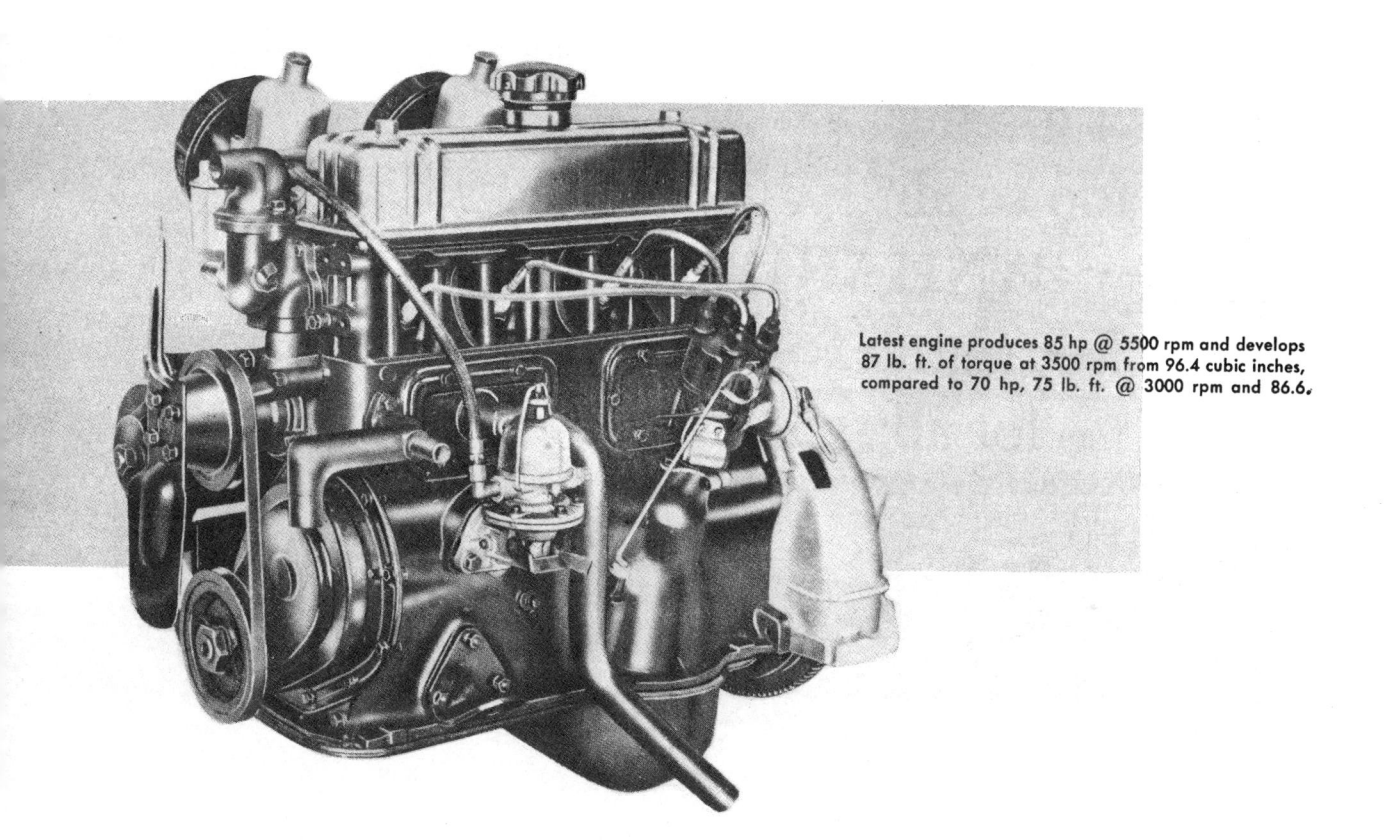

Latest engine produces 85 hp @ 5500 rpm and develops 87 lb. ft. of torque at 3500 rpm from 96.4 cubic inches, compared to 70 hp, 75 lb. ft. @ 3000 rpm and 86.6.

The jet sleeve, in which the fuel needle moves up and down, can also be moved up and down to provide a rich or lean mixture proportion independent of the size of the air throat opening. In other words, the richer mixture needed for starting from cold is initially provided by increasing the proportion of fuel to air (throat size). The dashboard choke control (fuel-enrichening) knob pulls down the jet sleeves in both carburetors. There is a cable running from the control to a tandem hookup to the jet sleeve levers at the base of the carburetor.

To further explain the principle, let's consider what happens when the engine is stopped. When engine fades after stopping, the pistons in both carburetor domes drop downward and virtually close the air passage and practically fully seat the tapered fuel needle in the fuel sleeve.

When starting, the use of the choke knob lowers the fuel sleeves and provides proportionally more fuel (or an enriched mixture) to the limited opening in the air passage. When the engine fires, vacuum builds up and the pistons in the dome rise to enlarge the size of the air passage and provide a greater amount of enriched mixture to handle the engine's increased breathing capacity or rising speed.

It should be pointed out that the dashboard choke has two functions: to provide a fast idle, and to operate the starting device or enrich the mixture. When you pull out the choke control, you will notice that it slides easily to begin with—this influences idling speed. If you pull it even further, it operates more stiffly as the engine is being choked through the lowering of the fuel sleeve.

When starting a cold engine, you pull out the choke control until increased resistance is felt and then a little more (to operate choke), depending on how cold the weather is. When the engine has run until the choke is no longer necesary, the button should be pushed in until it slides easily. At this point beyond the half way mark, the jet sleeves return upward to normal position and the button then only regulates the rolling speed. This is done by an idle speed cam disc on the side of the carburetor.

The car or engine should never be raced or violently accelerated when the knob is "out" in the choke position. If you do, you run the risk of the enriched mixture washing the oil off the cylinder bores and this may lead to piston scoring or seizure.

Continued on Page 94

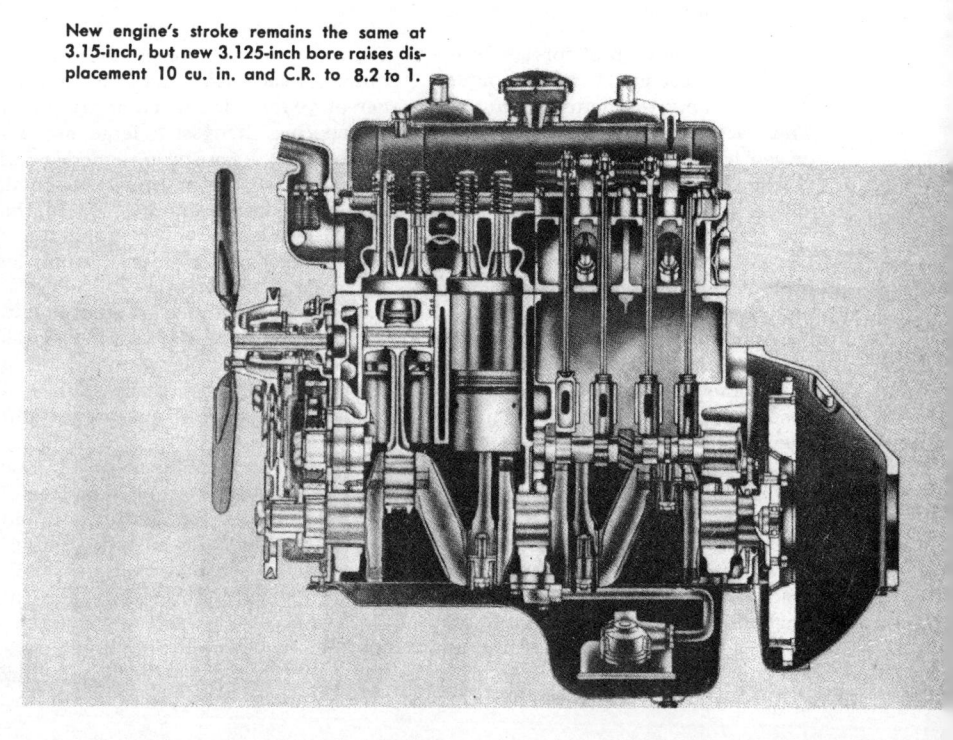

New engine's stroke remains the same at 3.15-inch, but new 3.125-inch bore raises displacement 10 cu. in. and C.R. to 8.2 to 1.

Volvo may look like copy of '46 Ford, but its performance is of a different order entirely.

TRACTABLE TIGER

Volvo calls its new PV 544 the "family sports car." A single drive will convince you the Swedes aren't saying near enough.

By AL BERGER

OF all the crop of foreign "economy" sedans, Volvo is unquestionably the hottest performer. This fact has been very thoroughly driven home at the past two 10-hour "Little Le Mans" endurance races, where Volvos easily captured the first five places in 1957 and the first three last year. Just to underscore it, however, Art Riley, who drove the winning car in the Little Le Mans, entered it in a number of regional SCCA races. More than once, he ran away from not only other sedans but full-fledged sports cars, winning one Lime Rock event against a field of 18 MGA's!

Before going farther, I might as well confess to a prejudice—I enjoy driving the four-speed Volvo more than any other sedan anywhere near its price class. Its high performance and precise handling, coupled with its small size, make it the best city-traffic machine I know of, and I live and work in New York.

Although Volvo follows the usual European automotive practice of continuing a successful model for a number of years without major styling or engineering changes, a large number of features have been added since SPEED AGE tested the three-speed, 70-bhp model of the PV 444 in the July, 1957 issue. For '59, Volvo made enough changes to justify changing the model designation to PV 544.

The 544 is not much different in outward appearance—the visible changes are a larger, one-piece curved windshield with slimmer corner pillars, a bigger rear window and larger tail lights. Inside there is a completely new dash and instrument panel, new suspended accelerator pedal, and safety belt anchors for both front and rear seats. The belts, an extra-cost option, are the sensible over-one-shoulder type which its Swedish sister car, the SAAB, also uses. The handbrake has been moved to the approved position between the front seats. Elbow rests recessed into the sides of the body make the rear seat roomier and more comfortable. In addition, there are a number of new color and upholstery combinations.

Other minor, but worthwhile, improvements include more powerful and larger direction indicators (hence the larger tail lights), and nozzles for windshield washers.

The power plant is the 85-hp version of the Volvo sports engine, which Volvo introduced in 1957. Completely conventional in concept, it is a four-cylinder in-line, water-cooled, push-rod-overhead-valve design, very similar to the MG engine. Bore and stroke are 3.125 x 3.15 in., displacement 97 cu. in. (1584 cc). The maximum power of 85 bhp comes in at 5,500 rpm, the 87 lb/ft. torque peak at 3,500 rpm, just where it's needed for passing bursts.

With the four-speed, all-synchromesh transmission to control it, the Volvo engine provides startling performance, which has been proved over

Four-cylinder, ohv engine has a 3.125 in. bore, 3.15 in. stroke. Twin SU horizontal carbs feed the 97 cu. in. mill which puts out 85 horses, 87 lbs/ft. of torque.

Completely new dash and instrument panel follows American styling. Round speedometer dial has been replaced by a horizontal thermometer version.

Interior is roomy and comfortable. Note the rear-seat recessed armrests, Headroom is adequate. The front seats fold back to form a two-place bed.

and over in winning more than 100 speed, economy and endurance contests all over the world. The long, floor-mounted shift lever operates smoothly and precisely. If the gearbox can be faulted at all, it is in the choice of gear ratios, which could be improved by closer spacing of second and third—if you allow the engine to wind up in second before shifting up, there's a noticeable drop in pickup while the revs build back up.

The old three-speed transmission is still available as an option, by the way, but I don't know who would order it except, perhaps, some Swedish immigrant who began driving Volvos in the old country when the first model came out in 1927, and is reluctant to learn the four-speed pattern.

Steering has been improved, not that it seemed to need it. An hourglass worm-and-sector gearing arrangement, with three turns lock-to-lock, provides quick, precise and accurate response without lag or backlash. It made short work of my favorite steering test, through the Grand Central Terminal's twisting elevated tunnel, whipping easily through the brutal up-and-down turns in third gear without any need for braking or downshifting.

This combination of power, maneuverability and the car's small size

There's room for plenty of luggage in trunk. Spare is easily accessible on right side.

make it an ideal traffic car. Just five feet, 2½ inches wide, it can slip through openings that larger cars must pass up, while the instant and enthusiastic response of those 85 horses will permit only the hottest of the domestic iron in the hands of the most alert drivers to out-drag it from a traffic light. As a result, you can make good time through rush-hour traffic that would hold you trapped and fuming in either a larger or less powerful car. Parking is a joy—the light, quick steering works the Volvo easily into any space a hand's breadth longer than its own compact 177 inches.

"Compact" is a good word to describe the Volvo's size, incidentally because it is the word George Romney of American Motors appropriated to describe the size of his Ramblers. The Volvo's lines are deceiving; it is not a small car, in the sense of the Volkswagen at all. It is only an inch shorter than the Rambler American, for instance, two inches *longer* than the Studebaker Lark, an inch longer than the Peugeot 403. There is ample room for five passengers, although four are more comfortable, and nearly 18 cubic feet of trunk space.

Volvo's boast, therefore, that their machine is a "family sports car," is no idle one. It has every attribute required for a comfortable, practical, tractable family car which the lady of the house can drive with ease. But it becomes a tiger when the competition-minded member of the family takes it out to the track. He can have a ball dicing not only with other sedans but with real two-seater, open-air sports cars, and stand an excellent chance of bringing home some silverware.

For rallying, too, the car is hard to beat. Designed to cope with Scandinavian terrain and weather, the Volvo is not likely to be handicapped by even the roughest road conditions. The all-welded, integral body-frame is rattle-free and is rust-proofed at the factory. The suspension is conventional, but sturdy and effective. The front wheels are independently hung with coil springs, control arms and double-action hydraulic shocks, the rear has coils with torque arms and track rod.

Braking is excellent, with light pedal pressure and very little dip. In our series of acceleration and braking runs, there was barely a trace of fade after a dozen hard stops, and once around the track to cool off was more than sufficient to restore their original power.

The same qualities that make it a good city car make the Volvo a fine cross-country car. The precise steering that finds holes in traffic handles winding country roads with equal ease, and the power that is so useful in the Grand Prix de Stop Light allows comfortable cruising in the 70-mph range.

Standard equipment also includes whitewall tubeless tires, dual sun visors, fresh-air heater-defroster, the safety-belt anchors and tool kit.

All in all, it's a hard-to-beat package at the price ($2330 P.O.E. New York), if you don't mind your friends' remarks that "It sure looks like a '46 Ford." It does, but no '46 Ford ever went like this one. But then, as I said, I'm prejudiced.

SPECIFICATIONS: VOLVO PV 544

ENGINE AND CHASSIS
- ENGINE .... B 16 B
- CYLINDER LAYOUT ...... 4 IN LINE
- BORE .... 3.125 INCHES
- STROKE .... 3.15 INCHES
- DISPLACEMENT .... 97 CU. IN.
- COMPRESSION RATIO .. 8.2:1
- VALVES .... OH PUSHROD
- CARBURETION .... TWIN SU HORIZONTAL
- TRANSMISSION .... FOUR-SPEED, FULLY SYNCH

OVERALL RATIOS
- 1st 15.7:1
- 2nd 9.93:1
- 3rd 5.97:1
- 4th 4.55:1
- REAR AXLE RATIO 4.55:1
- STEERING .... HOURGLASS WORM AND SECTOR

TURNS
- (LOCK TO LOCK) 3.2
- TURNING DIAMETER .... 36 FEET
- BRAKE LINING AREA .. 144 SQ. IN.
- SUSPENSION: FRONT .. IND. WITH COIL SPRINGS, CONTROL ARMS
- SUSPENSION: REAR .... IND. COIL WITH TORQUE ARMS, TRACK ROD
- WEIGHT .... 2,140 LBS.
- FUEL CAPACITY .... 9.5 GALLONS

DIMENSIONS
- OVERALL LENGTH .... 177 INCHES
- OVERALL WIDTH .... 62.5 INCHES
- OVERALL HEIGHT .... 60.25 INCHES
- WHEELBASE .... 102.5 INCHES
- TREAD: FRONT .... 51 INCHES
- TREAD: REAR .... 51.75 INCHES
- GROUND CLEARANCE .. 8 INCHES

PERFORMANCE
ACCELERATION THROUGH GEARS
- 0-30 MPH .... 4.1 SECONDS
- 0-40 MPH .... 6.6 SECONDS
- 0-50 MPH .... 9.6 SECONDS
- 0-60 MPH .... 12.9 SECONDS
- 0-70 MPH .... 18.6 SECONDS
- STANDING ¼-MILE ...... 18.9 SECONDS
- SPEED AT END OF ¼ .. 71 MPH
- MAXIMUM SPEED .... 95 MPH
- MAXIMUM OUTPUT   85 @ 5500
  (BHP @ RPM) ....
- MAXIMUM TORQUE   87 @ 3500
  (LBS/FT @ RPM) ......
- BHP PER CU. IN. .... .88
- LBS. PER BHP 25.1
- MILEAGE (ALL TESTS) .. 27 MPG
- PRICE $2,330

CT&T ROAD TEST

Volvo 60 h.p.

0-70 mph — 23.0 secs.
Top speed (mean of three runs) — 83 mph
Top speed through gears:
1st: 28 mph
2nd: 45 mph
3rd: 70 mph
Test Weight: 2,350 lbs.
Gear Ratios: (3-speed box not fitted on test car but is standard) 1st: 3.13:1; 2nd: 1.55:1; 3rd: 1:1; Rev.: 3.25:1.
Fuel Consumption Range: 26-32 mpg

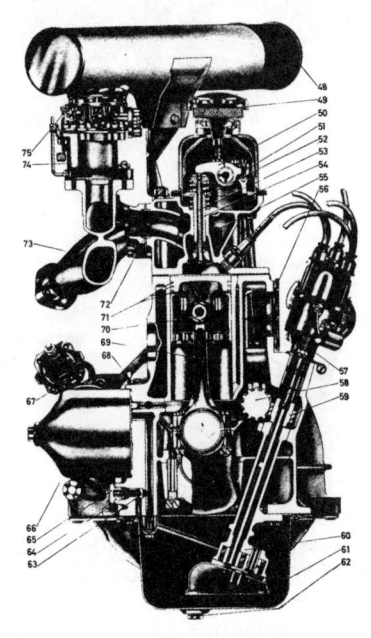

CANADA TRACK AND TRAFFIC has long been convinced that in the PV 544, Volvo markets one of the most acceptable, and best performing of all import sedans. The vigor with which these Swedish car perform has won the heart of many a car club enthusiast who has discovered that the handling, willing engine and high clearance makes the PV 544 an excellent rally car.

The Volvo is also readily adaptable for racing and was the outstanding marque in sedan races in Canada last year. Best of all, from the family man's point of view, the PV 544 provides ample room for four adults.

Now Volvo is offering a lower-cost version of the PV 544 which while yielding a little in performance gains on fuel consumption and is generally a more docile machine for day-to-day driving, commuting and touring.

The new PV 544, which can be distinguished from other models externally only by lack of chrome trim, is fitted with a 60 hp engine as against 85 hp in the more sporting version.

The 60 hp PV 544 mounts an engine of the same general design as other cars in the 544 range but is equipped with a milder cam and a single Zenith down-draught carburetor. The capacity of the engine, 1580 cc, is unchanged but the compression ratio is lowered from 8.2:1 to 7.4:1. The 60 hp model produces 85.4 ft/lbs of torque at 2,500 rpm as against 86.8 at 3500 in the 85 hp model.

The 60 hp Volvo is delivered equipped with a three-speed gearbox which is fully synchromesh on all forward gears, a rarity for three-speed boxes, and while the smaller box is adequate, a four-speed box, also fully synchromeshed is available at extra cost and permits the gentler engine to be used to its fullest extent.

The test car was equipped with four forward speeds but we found that the lowest gear was virtually unnecessary for level starts with two persons on board but the lower gear will be appreciated by those motoring in the mountainous country or carrying heavy loads.

Maximum speed reached in low was 26 mph; in second, 45 mph; in third, 69 mph. There is, however, a wide gap between second and third which is especially noticeable when it becomes necessary to change down in heavy traffic. A closer ratio here would be a definite advantage.

A glance at the performance table will show that the 60 hp version is quite a sprightly performer with a creditable 0-60 mph time of 16.2 secs but the power falloff at higher speeds is reflected in the extra 6.8 secs needed to reach 70 mph.

Engine noise was quite moderate at high speeds and the handling characteristics are of the same excellent nature as in the higher powered models. The interior trim is good quality, the chief difference between the 60 hp model and the more expensive PV 544 line being in the textile upholstery employed in the lower horsepower version.

The 60 hp model sells in Toronto with a three-speed gearbox for $2,095.

In all other respects the 60 hp models resemble exactly the 84 hp PV 544 already familiar to and popular with Canadians.

PERFORMANCE
Test Car: Volvo PV 544 60 hp
Type: Four-seat sedan, two-door
Engine: 4-cylinder, ovh watercooled

TIMES
0-30 mph — 5.4 secs.
0-40 mph — 7.9 secs.
0-50 mph — 11.1 secs.
0-60 mph — 16.2 secs.

51. Setscrew with lock nut
52. Valve spring
53. Valve
54. Valve guide
55. Water distribution pipe
56. Inspection cover
57. Distributor
58. Camshaft gear
59. Pump gear
60. Oil pump
61. Oil pan
62. Drain plug
63. Relief valve plunger
64. Relief valve spring
65. Nut and washer for relief valve
66. Oil cleaner
67. Starter motor solenoid
68. Water drain cock
69. Connecting rod
70. Piston pin
71. Piston rings
72. Guide sleeve
73. Exhaust manifold
74. Carburetor
75. Rubber gasket

A CAR ROAD TEST

A side view of the Volvo "Sport" shows up the clean lines which, though appearing slightly old-fashioned, are in fact most efficient aerodynamically.

Volvo PV544 "Sport" Saloon

THE Swedish Volvo is a comparatively new name in South Africa's motor world, yet in the few years it has been on the scene it has established a very high reputation for sturdy dependability allied to good performance. The achievements of the high-performance "Sport" version in rallies and on the track have shaken not a few people, so that we welcomed the chance to conduct a thorough test of the latest version of this model.

Outwardly it looks the same as the previous version, with distinctly old-fashioned but efficient lines, the changes lying mainly in the gearbox and front suspension.

In the new M40 gearbox, which also has synchromesh on all four forward ratios, the gaps between first and second and between second and third have been narrowed, and that between third and top has been widened. The result is closer spacing of the ratios overall, and the car should theoretically have slightly poorer acceleration but a higher maximum in first and second and a lower maximum but better acceleration in third.

At the front end the former single damper for each wheel has given way to twin "shockers" on each side. Air-filled rubber bellows, with a controlled rate of leakage, replace the old rubber bump-stops to provide progressive stiffness.

The 1,580 cc engine is "just one of those things". With its vertical, in-line pushrod o.h.v. it is seemingly quite ordinary, if not also old-fashioned, yet it produces the by-no-means ordinary figures of 85 bhp and 87 lb-ft of torque (SAE) at 5,500 and 3,500 rpm respectively, and runs very happily up to 6,000 rpm. Which helps explain how the car is happy to pull a 4.1:1 rear axle ratio and 5.90 x 15 tyres.

The car had covered about 2,500 miles when it was handed over and, though it settled down to cruise quite happily and steadily at 80 mph, it definitely lacked the "edge" to give what it should have done when it came to making the timed runs. Stiffness and not poor tuning was quite clearly to blame, so these runs were repeated some weeks later when the car had covered 5,100 miles.

The figures then recorded speak for themselves. Bare figures, however impressive or otherwise, in any case form only one of the chapters in the story of any good car, and it was in the manner of its going that we particularly enjoyed the "Sport".

Even while the engine was still stiff there was quite enough acceleration available in third and top to nip past other fast-moving traffic very quickly. But this is no excuse for being lazy about using the gearbox.

This is first-class. It is operated by a rather long but very rigid lever, with a man-sized knob, in the proper place on the floor. The synchro proved quite unbeatable, and the only criticisms were that the spring-loading of the lever towards third and top was slightly too strong, and that in top the knob was rather close to the passenger's left leg.

This could lead to embarassing misunderstandings. And if, to avoid them or the chance of being bruised by a driver who is swopping cogs in a hurry, the passenger shifts his (her) feet, they are likely to end up on a portion of the floor which is very hot indeed as a result of the too-close proximity of one of the two mufflers in the exhaust system.

About the steering there is little to be said. It is light and positive, reasonably high-geared and completely free of lost motion or sponginess.

As far as handling and roadholding are concerned, the "Sport" is comfortably stable. Understeer is definite but not excessive — though the noises made by the tyres of non-standard tread pattern on the test car seemed to be calling all the traffic cops in creation — and roll is well restrained.

Location of the coil-sprung rear axle is good, but could definitely be better. There is sufficient give in the combination of radius rods and Panhard rod which take care of locating the axle to allow it to make noticeable and unwanted contributions to the steering, particularly when the car meets diagonal bumps on a fast bend. This can be disconcerting until experience shows that it does not lead to signs of the tail breaking away or other bad reactions even when the car is very close to the limit.

The combination of the car's general handling characteristics and a third gear in which well over 70 is available with no sweat make a twisty road a challenge which it is a delight to meet.

There is a section of 100-odd miles of a certain national road, well-known to thousands of drivers, which offers just about everything that is needed thoroughly to sort out every facet of a car's performance. Not much of it is particularly level and some of the gradients are distinctly tough, but there are smooth fast stretches with sweeping bends, towns to negotiate and 40-some miles of narrow, twisty, bumpy up-and-down in-and-out stuff where brakes, steering, gearbox, acceleration and roadholding are all severely tried if a fair average is maintained. It is also a road which carries heavy traffic and, in the nature of things, most of it is always to be found travelling slowly at the most awkward spots.

Over this stretch the Volvo averaged 58.35 mph going down and 59.44 mph going up and frightened nobody — we hope. Fuel consumption is normally also checked over this stretch, to give an indication of the lowest sort of figure an owner can expect when driving hard on the open road, but in the case of the Volvo the stiffness of the engine again made itself felt, so this figure too was established later.

The car's brakes proved adequate to cope with the rest of its performance, even on this stretch, but were

The neat and uncluttered theme of the car is continued in the "cockpit". The very sturdy gear lever and size and spacing of the pedals will be noted.

The boot of the "Sport" has a very generous capacity indeed, and the sporting young man with a family will find it possible to get the pram into it when necessary.

From the front the "Sport" is tidy and uncluttered. The Swedes, though they also drive on the left, consider left-hand drive cars safer, and there is no news of a right-hand drive version.

69

The front-hinged bonnet is not only a safety feature, but lifts far enough to make everything under it, including the battery, very accessible. It will be noted that only one low-tension lead to the coil is to be seen.

not outstanding. They tended to pull to one side after repeated hard use at short intervals, and pedal pressures were rather high, but they did not fade. The handbrake was very good, with a solid pull-up lever pivoting between the front seats, and a neat guard to prevent accidental releasing of the catch.

The ride over a stretch of rough dirt road was distinctly lively, probably aggravated by tyre pressures set for fast going and the fact that we were only two up. No dust penetrated the body and only the slightest trace could be found in the boot, but heavy rain disclosed a leak in the scuttle over the passenger's legs.

Generally the car proved very comfortable, with ample leg-room in front and more in the rear than the figures would lead one to expect, thanks to clever design of the bases of the front seats which makes all the room usable and also makes access to the rear seats easier than in most two-door cars. The driving position is good, the front seats — also altered in this version — being well shaped to support their occupants against sideways forces and having a good range of fore-and-aft adjustment, though the rake of the backrests is only adjustable when the car is stopped and the seats unoccupied.

Visibility forward and to the sides is good, but to the rear is restricted by the small effective depth of the rear window. The rear-view mirror should be slightly lower. Because of the body shape the near-side front wing cannot be seen from the driving seat, but this is of little moment, as the fact that the car has left-hand drive makes accurate positioning relative to the roadside very easy.

Relative distances from the seat to the wheel and pedals have been well judged, and the pedals themselves are good as regards both size and spacing, so that it is easy to "heel-and-toe" yet not even the largest feet should find themselves pressing both brake and accelerator inadvertently. A neat touch is the pivoting of the pendent accelerator pedal so that it automatically adjusts itself to the angle of the driver's foot.

The instruments are grouped directly in front of the driver under a slight hood which is enough to prevent reflections in the screen. They consist of a ribbon-type

From the rear quarters the "Sport" presents a view of a long sloping back — which Americans used to call a "fast back". The rear quarter windows hinge outwards at their trailing edges for ventilation, and the single door on each side is large enough to make getting into and out of the rear seats easy.

speedometer, which could be steadier, total mileage recorder, "trip" recorder reading in tenths and which can be zeroed, water temperature and fuel contents gauges with vague markings, and lights for oil pressure, ignition, high-beam and to show that the indicators are operating. Lifting the end of the "stalk", which operates the indicators, towards the wheel flashes the headlamps.

The lack of an oil pressure gauge and ammeter are to be regretted. Even more to be regretted is the lack of a tachometer, which would be both very useful and well in keeping with the car's character and performance.

Minor controls and switches are well positioned, positive in operation and clearly labelled. The push-pull headlamps switch incorporates a rheostat to control the intensity of the instrument lighting.

Finish is excellent, outside and in. The plastics material used to cover the seats has "breathing" holes, the dash top is padded with non-reflecting black material, and the floors are covered by sensible rubber mats, with underfelting attached. A central ashtray and lighter are provided in front, and two ashtrays in the rear. The glove-box has a locking lid which greatly increases its usefulness, but there is no other accommodation for oddments except a small shelf behind the rear seat. The door "keeps" are poor.

70

SPECIFICATION

MAKE AND MODEL: Volvo PV544 "Sport".

ENGINE: 4-cylinder, in-line, water-cooled, pushrod o.h.v., twin HD4 (1½ in.) S.U. carburettors, compression ratio 8·2 : 1.

BORE AND STROKE: 79·37 x 80 mm. (3·125 x 3·15 in.).

CUBIC CAPACITY: 1,580 c.c. (98 cu. in.).

MAXIMUM HORSEPOWER: 85 b.h.p. (S.A.E.) (76 DIN.) at 5,500 r.p.m.

MAXIMUM TORQUE: 87 lb.-ft. at 3,500 r.p.m. (S.A.E.) (83·2 lb.-ft. at 3,300 r.p.m. DIN.).

TOP GEAR M.P.H. AT 1,000 R.P.M.: 18·46.

PISTON SPEED AT MAXIMUM H.P.: 2,880 ft./min.

BRAKES: Self-centring, duo-servo hydraulic. Total friction area 157 sq. in.

SUSPENSION: (Front) Independent, wishbone and coil, anti-roll bar. Twin shock absorbers and progressive-rate rubber-air cushion each side. (Rear) Live rear axle, located by rubber-bushed radius rods and Panhard rod. Coil springs.

TRANSMISSION: Single dry plate clutch, mechanically operated. Four-speed, all-synchromesh gearbox, manually operated by floor-mounted lever.

OVERALL GEAR RATIOS: 1st: 12·833 2nd: 8·159 3rd: 5·576 Top: 4·100 Rev.: 13·325

TYPE AND RATIO OF FINAL DRIVE: Hypoid, 4·10 : 1.

TYRE SIZE: 5·90 x 15.

LENGTH: 13 ft. 1 in. **WIDTH:** 5 ft. 2¼ ins. **HEIGHT:** 5 ft. 1⅛ ins. **WHEELBASE:** 102½ ins.

TRACK: (Front) 51 ins., (Rear) 51½ ins. **GROUND CLEARANCE (laden):** 7⅜ ins.

STEERING: ZF worm and roller, 3¼ turns lock-to-lock. **TURNING CIRCLE:** 35½ ft.

FUEL CAPACITY: 7·75 galls. **OIL CAPACITY:** 6¼ pints (including filter). **BOOT CAPACITY:** 17 cu. ft.

LICENSING WEIGHT: 2,140 lbs. **WEIGHT AS TESTED:** 2,440 lbs.

PRICE AT S.A. COAST: R1,920. **PRICE IN JOHANNESBURG:** R1,960.

INTERIOR DIMENSIONS:
Width of front seat(s): 43⅛ ins.
Driver's seat to clutch pedal: (Max.) 21½ ins., (Min.) 16½ ins.
Front seat headroom: 3¾ ins.
Width of rear seat(s): 51¼ ins.
Rear seat kneeroom: (Max.) 10½ ins., (Min.) 5½ ins.
Rear seat headroom: 1½ ins.
(Seat widths measured between arm-rests, headroom with 6ft. man seated, no hat).

PERFORMANCE

ACCELERATION THROUGH GEARS:

M.P.H.	Secs.	M.P.H.	Secs.	M.P.H.	Secs.
0–30	5·2	0–60	16·8	0–90	—
0–40	8·0	0–70	24·0		
0–50	11·65	0–80	39·15		

ACCELERATION IN HIGHER RATIOS IN SECONDS:

M.P.H.	Top	3rd	M.P.H.	Top	3rd
20–40	—	7·7	50–70	19·3	12·8
30–50	13·7	8·8	60–80	27·5	18·25
40–60	15·0	9·75	70–90	—	—

STANDING QUARTER MILE: 21·15 secs.

REASONABLE MAXIMUM SPEEDS IN GEARS:
1st: 37 m.p.h. 2nd: 60 m.p.h. 3rd: 81 m.p.h.

MAXIMUM SPEED IN TOP: 92·31 m.p.h.

MAXIMUM PULL IN GEARS:

	Lbs./Ton	Equivalent Gradient
1st:	584	1 in 3·71
2nd:	418	1 in 5·26
3rd:	290	1 in 7·7
Top:	178	1 in 12·5

BRAKING: Emergency stopping distance with car in neutral at 30 m.p.h.: 33·5 ft.

FUEL CONSUMPTION: 34·99 m.p.g. at an average speed of 52·14 m.p.h. over 118·1 miles.

TEST CONDITIONS: 4,200 ft., fine, warm, no wind, dry tarmac, 93 octane fuel.

(Cars supplied by New Curzon (City) (Pty.) Ltd., Johannesburg, and (for photographs) Louis Joss Motors (Pty.) Ltd., Pretoria.)

Accessibility under the bonnet, which hinges at the front for safety and must be both released and locked from inside the car, is good. Points worthy of note are that the fuel pump has a hand-priming lever, and that only one low-tension lead to the coil is visible — the other is enclosed in a heavily-armoured cable from the base of the coil, inside the car, to the ignition switch, which earths it and makes the car difficult to steal.

The boot is not only surprisingly large, but high enough to take awkward things like prams. The spare wheel is mounted vertically on the right side, where it is out of the way, and the floor is covered by a rubber mat. Even more space for goods can be quickly obtained by removing the back of the rear seat, which is held by clips.

Running with the front windows and ventilating panels closed, and relying on the (extra) fresh-air system and hinging rear windows to provide ventilation, the car was very quiet even at high speed. It became very noisy with wind roar as soon as it became necessary to spoil the good aerodynamic shape by opening something at the front, however.

Useful "extras"

Useful "extras" fitted to the test car, in addition to the fresh-air duct, included dual wing mirrors, a radiator blind and a locking filler cap, though the latter did bring up to three the total number of keys needed to operate all the locks on the car.

It was surprising in a car of this class, however, to find that the windscreen washer was also an "extra", though the discharge nozzles and connections are standard.

Also standard are the attachment points for safety belts, built in to reinforced portions of the body. The belts themselves are "extras", but nearer to being essential than merely useful these days. Those fitted to the test car were the latest type, combining a diagonal "sash" and a lap strap. They proved quite comfortable, irrespective of the sex of the wearer, and represented a good compromise between a full harness and a simple lap strap.

To sum up, the Volvo people, with their "Sport", have achieved a very good compromise between a reasonably economical family tourer, with room for four people and ample luggage in comfort, and a sports saloon of high performance. It is very good value for money in either guise. ●

A three-quarter front view of the "Sport"; a car representing a very successful compromise between a high-performance sports car and a family saloon with ample room.

ROAD TEST/11-16

VOLVO 544 SPORTS

Volvo has another winner — the customers keep proving it!

THE VERY UPRIGHT VOLVO TWO-DOOR has been around for several years now with only minor changes in its basic configuration. One of those rare cars that almost defies criticism by experts — with the possible exception of the "pre-war" body design — it has gradually been improved and updated within its rugged skin. The latest of these improvements is the addition of the 1780 cc engine similar to the one installed in the 122S that we tested in the last issue of SCG. On this option Volvo hangs their "Sport" tag, installs the deluxe vinyl upholstery, a nametag on the rear deck lid and a "B-18" designate in the front grille. The last of these indicates the new powerplant and, more specifically, this is the B-18-D version. For clarification between the "D" used in the 544 and 122S, and the "B" in the P-1800, we break them down this way:

B-18-D	B-18-B
90 hp @ 5000	100 hp @ 5500
105 lb/ft. torque	110 lb/ft. torque
8.5 to 1 comp. ratio	9.5 to 1 comp. ratio

Other differences include a more conservative cam timing for the "D" and aluminum bearings for the mains, instead of lead-bronze for both mains and rods as is included in the "B". Both are rugged, compact units with five-main-bearing crankshafts, full-flow filters (the "B" has an oil cooler as well), machined combustion chambers and full water-jacketing around the spark plugs. Light alloy is used in the bellhousing and timing cover.

Getting back to the chassis, it's been over two years now since we've driven a Volvo Sport and it became quickly apparent that many effective changes had been made. First off, the deluxe interior of this model is refreshing and tastefully done. We'd forgotten how straight up you sit and, although it's weird at first, the position is very comfortable and allows for good control. The steering wheel is fairly close but it's positioned low and allows a relaxed arm position, with the left elbow resting on the armrest of the door. The foot pedals are spaced nicely with the clutch and brake floor-mounted, the gas pedal hanging. The latter, at least in our test car, was a bit touchy for smooth feed at low rpms and we suspect that some of this was due to poor geometry in the linkage actuating the dual carburetors.

The dash controls for choke, lights, ignition, and wipers are all very nicely placed and contribute to the overall ease of operation. Steering pressure is at all times very light and the car almost overly responsive. The rugged, floor-mounted shift operates smoothly through the gears of the full-synchro, four-speed transmission. Ratios in the latter are evenly spaced to use the engine's wide torque range and are somewhat closer in ratio than those in the first four-speed transmission that Volvo produced.

Except for fair-sized blind spots in the rear quarters — this is magnified by the height of the car — visibility is excellent. Possibly a tall person would find that the upper edge of the windshield was too low; it was certainly unusual for us to be looking directly out the upper half. This

Styling of the 544 remains as standardized as VW. Its solid and efficient construction more than offsets this, however, and has helped Volvo to build an equally solid reputation for a reliable product. Volvo is now the fourth largest car importer.

PHOTOS: RANDY HOLT

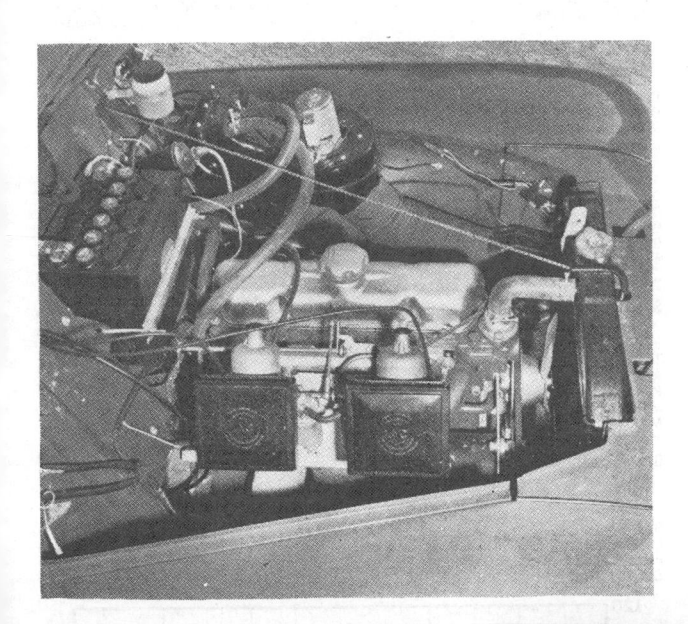

The Sport engine is a warmed-up version of the Volvo workhorse that delivers a relaxed 90 horsepower, moves car along well, enhances good handling.

Interior is up-to-date with padded dash and Detroit-like instrumentation. The choke is extreme left knob. All controls are centralized around the large wheel.

Uneffected by almost any type surface or terrain, the Volvo soaks up abuse and charges on. Choppy bumps are felt but stability remains high. The general ride characteristics are above average.

VOLVO 544

provides an excellent view of the road directly ahead of the short hood and should keep a very short person from having to peek through the steering wheel.

The normal-road ride of the 544 is very likeable — both stable and relaxed. Tar strips are felt, however, and small, sharp bumps will make the rear end go hippity-hoppity very easily. In a choppy corner this considerably increases the slight oversteer but, in a smooth corner, the car goes around like a Formula Junior — very rapid and dead smooth.

Handling is very definitely the forte of the Volvo 544. It can be, and was, whipped around mountain roads at speeds that would have embarrassed many a first-rate sports car. The combination of engine, gearbox, and brakes allowed it to take all kinds of grades and corners in complete stride with a minimum of effort required from the driver, much like giving a good horse its head in rough going. We were frustrated during our first outing with a photographer in attempting to obtain action shots. We flung the car through some corners at a *very* brisk clip but the resultant photos made the car look as though it were being calmly motored around. On the second trip we made *sure* the action looked right, but it's a chore to get the high sedan "out of shape".

The noise level is considerably improved over previous models. Some resonance is to be expected in a solid, unitized body, but it's held pretty much to a minimum. With the windows closed tight, normal conversation can be held at very high speeds. Opening any of the windows at normal turnpike speed will produce a considerable roar, however. This is due to the boxy frontal area of the sedan and it's doubtful that anything other than extensive restyling would alter it.

Quite pleasing were all the access openings — doors, hood, and rear deck — with extremely logical hardware and hinges. The hood opens to full vertical on counterbalanced hinges and allows a maximum area in which to service the engine compartment. The trunk lid is likewise counterbalanced and exposes a small cave in which to store baggage of almost any size and shape. The doors, at least in a new unit, require some effort to close, but spring open at a touch of the handle.

Heating and defrosting, as would seem logical in a car with Scandinavian origin, are the best. We understand that blower capacity has been increased this year and it certainly turns up a storm in Hi position. The chain-operated shroud in front of the radiator still governs the engine's coolant temperature and, even with the shroud pulled completely closed, it takes about two miles to get the temperature gauge up to the normal. Once this happens, however, there's heat a'plenty for warming the interior in rapid order. Radiator size has likewise been increased with installation of the B-18 engine and we found it necessary, even on warm days, to run with the shroud partially closed to get the engine up to proper temperature. A proper warm-up makes for a noticeable increase in gas-mileage figures, too. We found the all-around average to be about 24 mpg; impressive in view of the kind of driving we were doing and the fact that there was more than adequate performance available at all times.

The Volvo 544 has been summed up as a cross between a tank and a sports car. It's even more of the latter with the new engine. Very utilitarian in every respect, it sacrifices nothing in the way of being fun to drive, tolerating all kinds of abuse in the process. There seems little doubt that this same basic model will be around for several years more. The company knows it has a winner and its customers keep proving it at every chance.

— *Jerry Titus*

ROAD TEST/11-62
TEST DATA

VEHICLE ....Volvo   MODEL ....P544 Sport
PRICE (as tested) ....$2395 POE L.A.   OPTIONS ....None

ENGINE:
Type: ....B-18D, 4 cycle, 4 cylinder, in-line, 5-main
Head: ....4 port, removable
Valves: ....OHV pushrod/rocker
Max. bhp ....90 @ 5000 rpm
Max. Torque ....105 lbs. ft. @ 3500 rpm
Bore ....3.313 in. 84 mm.
Stroke ....3.15 in. 80 mm.
Displacement ....109 cu. in. 1780 cc.
Compression Ratio ....8.5 to 1
Induction System: ....2 sidedraft SU carburetors
Exhaust System: ....cast header single pipe
Electrical System: ....12V distrib. ignition

CLUTCH:
Single disc, dry — Diameter: ....8½ in.
Actuation: ....mechanical

TRANSMISSION: ......M40 4-speed,   **STEERING:**
                      full synchro   Turns Lock to Lock: 3¼
Ratios: 1st ....3.13 to 1   Turn Circle: ....32 ft.
        2nd ....1.99 to 1   **BRAKES:**
        3rd ....1.36 to 1   Drum or Disc Diameter ....9 in.
        4th ....1.0 to 1   Swept Area ....208 sq. in.

DIFFERENTIAL: ....Hypoid
Ratio: ....4.1 to 1
Drive Axles (type): ....enclosed, semi-floating

CHASSIS:
Frame and Body: ....Unitized steel with subframes
Front Suspension: ....Unequal A's, coil springs, tube shocks, anti-sway bar
Rear Suspension: ....live, coils, trailing arms, tube shocks
Tire Size & Type: ....5.90 x 15 Goodyear

WEIGHTS AND MEASURES:
Wheelbase: ....102.4 in.   Ground Clearance ....7.8 in.
Front Track: ....51 in.   Curb Weight ....2354 lbs.
Rear Track: ....51.75 in.   Test Weight ....2624 lbs.
Overall Height ....61.3 in.   Crankcase ....3 qts.
Overall Width ....6.26 n.   Cooling System ....4 qts.
Overall Length ....17.5 in.   Gas Tank ....9 gals.

PERFORMANCE:
0-30 ....3.3 sec.   0-70 ....20.7 sec.
0-40 ....5.9 sec.   0-80 ....30.4 sec.
0-50 ....9.3 sec.   0-90 ....— sec.
0-60 ....13.8 sec.   0-100 ....— sec.
Standing ¼ mile ....18.9 sec. @ 68 mph
Top Speed (av. two-way run) ....93 mph
Speed Error....30   40   50   60   70   80   90
Actual.........30   40   50   60   69   79   88
Fuel Consumption   Speed Ranges in gears:
Test: ....24.5 mpg   1st ....0 to 28 mph
Average: ....27 mpg   2nd ....10 to 49 mph
Recommended Shift Points   3rd ....20 to 68 mph
Max. 1st ....28 mph   4th ....30 to top mph
Max. 2nd ....49 mph
Max. 3rd ....68 mph
Brake Test: ....70 Average % G, over 10 stops
Fade encountered on 9th stop.

REFERENCE FACTORS:
BHP per Cubic Inch 0.83
Lbs. per bhp 26.2
Piston Speed @ Peak rpm ....2620 ft./min.
Sq. In. Swept Brake area per Lb. 0.087

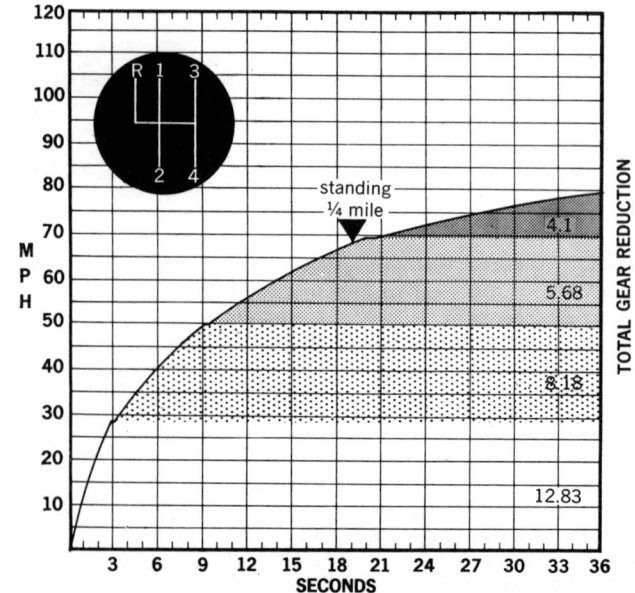

VOLVO 544

☐ In spite of its height, we've overlooked the ancient Volvo 544 sedan. This is easy to do, because of the attention the newer 122S, or Canadian, has attracted since its recent development as a two-door, a station wagon and its Canadian assembly. But our readers wouldn't overlook the 544. They've asked us constantly to do a road test on the new B-18 engined version. And the buyers don't neglect this remarkable, almost vintage machine. Why would anyone want the dated, "old Ford"-looking 544 at a price of over $2,700 when they can buy a sleek new compact for less, or the modern Canadian for a little more? We wondered too, until we tried one and got the surprise of our lives. The Volvo 544 is as mechanically modern, if not more so than any of the compacts. Its performance is little short of sensational, the construction remarkably rugged, the appointments attractive. The 544 with its new five-bearing crank engine will seemingly last forever and the resale value remains high. A high-performance engine, 4-speed all-synchro gearbox, heavy-duty suspension, bucket seats and whitewalls, are standard equipment!

coachwork

You don't stoop to enter a 544 — height has its advantages! Somehow this car has a form of reverse snob appeal. VW had it, but became too common. With umpteen different versions of North American cars all looking the same, Volvo's 544 looks down from its lofty height and says: "So what — it's fun to be different!" And that old Ford it reminds you of had tremendous appeal as a light, quick, sporting sedan. The 544 has this appeal too, in up-to-date mechanical form with a better-looking body. Vision to the rear suffers; so does tracking in high winds.

interior

As you might expect, there's plenty of headroom inside the 544. Legroom is spacious, front and rear. Elbow room is adequate, but the body shape restricts this to some extent and the separate fenders narrow the foot wells so that the left foot sits over the dip switch. Moving the dipper to the steering column would alleviate this. The seats are bucket-type, and feature leatherette upholstery with tiny holes, causing a flow of air which aids in summer cooling. The rear seat is exceptionally comfortable, with built-in arm rests at the side, and a large shelf at the back. However, we're less enthusiastic about the front seats. They're much too short (front to back) and though the back rests can be adjusted by removing shims, the rake is still insufficient for some drivers. We had the feeling we were going to fall off the seat at times. Appointments are complete: including locking glove box, arm rests, ash trays, padded dash and sun visors. Material quality is excellent.

instruments

Behind a nicely-styled steering wheel with horn ring sit the familiar Volvo instrument group. We can't help but wish it consisted of round dials but the car industry seems devoted to the horizontal speedometer and Volvo is no different. The speedo is the ribbon-type. Other instruments include fuel gauge, temperature, warning lights for oil an damps, plus separate mileage and trip odometers. The heater, which is ideal for tough Canadian conditions, is controlled by a simple system of levers in the upper left of the dash, aided by a rad blind adjustable from within, and a two-speed fan. The hand brake is between the seats, all pedals are large and well spaced. Push-pull knobs operate easily and there is an efficient hand choke.

engine

Many a tribute has been written about Volvo's B-18 engine since it first appeared. Undoubtedly this is the finest production four-cylinder in any sedan today. It is overhead valve, oversquare, with a five bearing crank, and according to reports in magazines all over the world, is virtually unbreakable in normal use. The 544 and 122 models have 90 horses, from 1800 c.c.'s with twin S.U.'s. In the P1800 the power is up to 108 and the marine version is rated at 110. (Oh, but to have THAT in the sedan!) The joy of the unit is that durability, power and economy are combined in one. Mated to it is a four-speed, all-synchro gearbox with well-spaced ratios. Changes are accomplished with a long lever, smoothly, though the thing looks as though it came from a 10-ton truck. The P1800 remote shift would be welcome.

trunk

You may get a pleasant surprise if you crawl into the 544's trunk. It is remarkably large and well able to cope with awkward objects. The vertically mounted spare is easy to remove. A folded baby carriage will fit lengthways into the trunk, with plenty of room left to pile on. We mention this because of the 544's appeal to family men with sports car hankerings; the ability to carry carriages, groceries, luggage and pottys all at one time is important to this group. Volvo's 544 qualifies well.

handling

Being accustomed to wheeling a Volvo Canadian, we found the 544 great fun to drive. To compare: the steering is lighter and the car feels much more agile. Obvious conclusion is that the 544 is smaller. Imagine our amazement when we checked and discovered that the 544 and the 122 are identical in length and wheelbase! Weight makes the difference — the 544 is considerably lighter, also narrower. Suspension system is well designed, with coils all-round. A stabilizer aids in the front while the rear utilizes torque rods and a track bar to keep the wheels properly located under stress. This, plus balanced weight distribution makes the 544 handle in the sports car tradition, one reason why these sedans have always amazed on the race track. There's a trace of understeer but we broke the tail away with ease, though always in control. Ride characteristics are pleasant. It is in the handling category that the title "Sports" seems most apt.

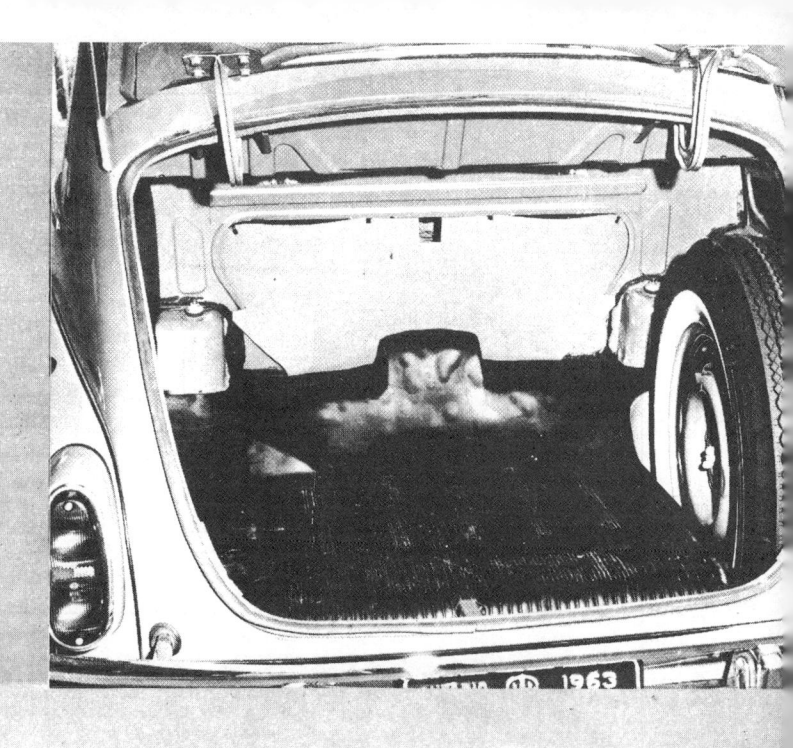

performance

Hold your hats, guys, this cat's got scat! No kidding, the B-18 makes Volvo's 544 into a real bomb. Acceleration is far superior to any six-cylinder compact and even the V-8's have to open the pot to keep up. But the real blessing in this performance is that fuel economy remains in the 25-35 miles per gallon category. Acceleration is constant right up to 80 miles an hour — after that the wind resistance takes over. Gear ratios seem stepped right for performance. High speeds in all gears are available but the low-speed lugging ability still permits top gear tooling in town. It isn't necessary to change down for ample passing power, though the sporting driver will enjoy extra go by making full use of the cogs. The brakes are not disc, nevertheless they didn't fade and stopped quickly. They weren't set up properly in our test car; high speed panic stops caused frantic wheel-twisting to keep a straight line.

summary

There's no doubt the legendary 544 still has lots of life yet. All that's old is the body style and even that, as mentioned earlier, has its own peculiar charm. Volvo calls the 544 the "Sport". We agree completely. The family man waiting, as we are, for a reasonably priced 4-seater GT car, might do well to check the 544. It has all the attributes of a GT except style and there's no doubt the continued shape helps retain resale value. Price isn't low, but as we pointed out, you get a lot for your money and the particular virtues of the 544 aren't common to lower-cost cars. The 544 is high, vintage-styled, technically advanced, comfortable, fast, sporting, rugged, economical, roomy. Come to think of it, it may even be a bargain.

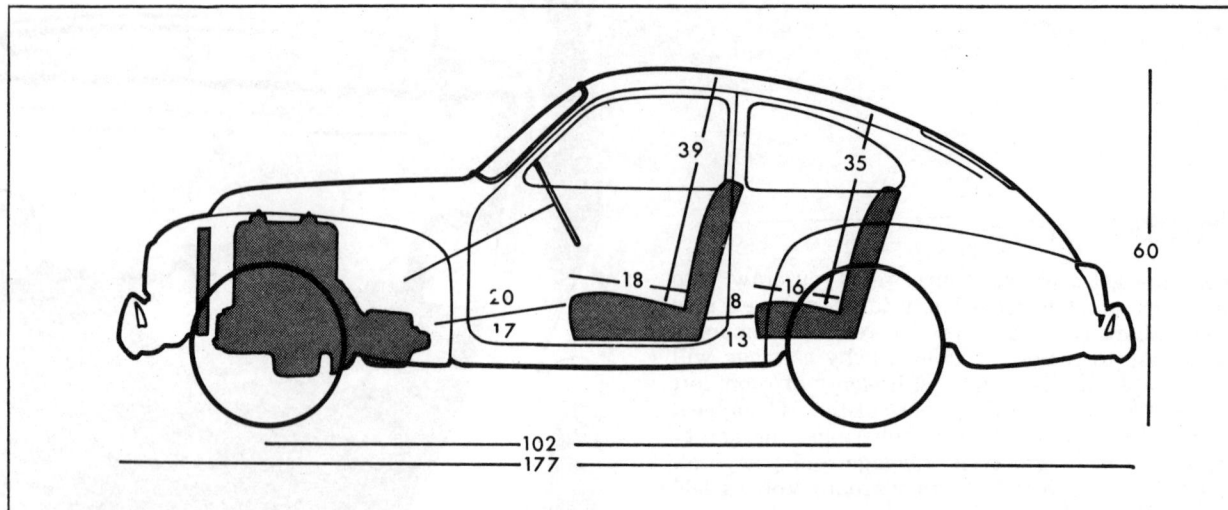

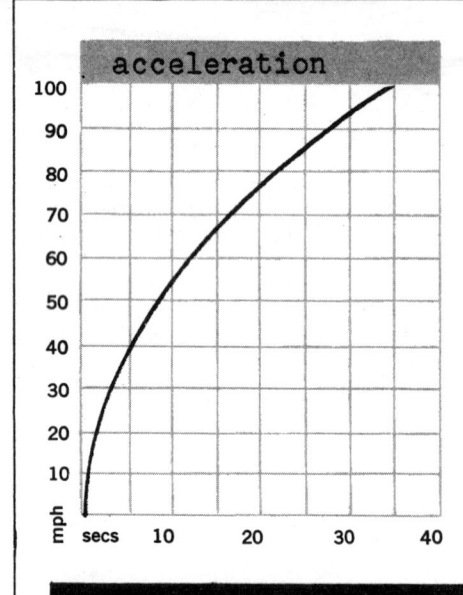

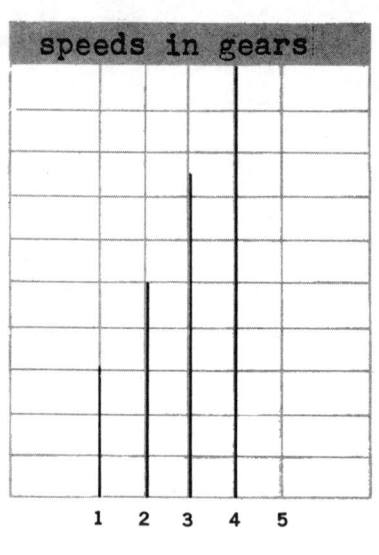

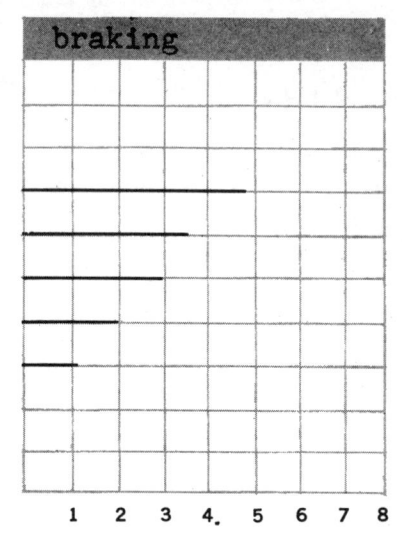

VOLVO 544

Engine:	Four cylinders in line, o.h.v., water-cooled
Bore:	3.313"
Stroke:	3.15"
Displacement:	1,783 c.c.
Compression ratio:	8.5:1
Maximum power:	90 h.p. @ 5,000 r.p.m.
Maximum torque:	135 lb. ft. @ 3,500 r.p.m.
Brakes:	Duo-servo drum, 139.8 sq. in.
Transmission:	4-speed, all-synchro
Ratios:	1st: 3.13:1; 2nd: 1.99:1; 3rd: 1.36:1; 4th: 1.1
Front suspension:	independent, coil springs, control arms, stabilizer
Rear suspension:	fixed diagonal support arms and torque rods, track bar, coil springs
Steering:	cam and roller, 3¼ turns l to l
Wheelbase:	102½"
Length:	175"
Width:	62½"
Height:	61½"
Weight:	2,220 lb.
Fuel capacity:	7.4 imp. gal.

TEST CAR COURTESY

VOLVO (CANADA) LTD.

acceleration times

0-30	—	3.0 seconds
0-40	—	5.4 seconds
0-50	—	7.9 seconds
0-60	—	11.0 seconds
0-70	—	15.4 seconds
0-80	—	22.5 seconds

SPEEDS
1st: 29; 2nd: 48; 3rd: 72; 4th: 98

R&T ROAD TEST

VOLVO PV-544

*If thou shouldst lay up even a little upon a little,
and shouldst do this often,
soon would even this become great.—Hesiod, 720 B.C.*

A TIDAL WAVE of cars from across the Atlantic came to our shores about 7 years ago, and in the full flowering of American enthusiasm for the imports, it was difficult to spot those destined to survive. However, few people would have selected the Volvo, a scaled-down 1948 Ford with a smallish 4-cyl engine, as a likely candidate. The Volvo was too dated in appearance, and embodied little in interesting technical features. Far-out engineering was getting most of the play in those days (as is often true now) and the completely straightforward Volvo drew little notice. Nevertheless, after the tides receded in 1960, Volvo was among those which had found a solid and satisfied following of American buyers.

In its original form, the Volvo had the same "1948 Ford bodywork" of the present 544, but was somewhat different mechanically. The early model had a 3-speed transmission and a sports-tuned version of Volvo's old PV-series 3-mainbearing engine, which had bore and stroke dimensions of 2.95 and 3.15 in., and a displacement of 1414 cc.

The new engine, carrying the designation "B-18," also has 4 cyl and a stroke of 3.15 in., just like its immediate ancestor, but that is where the resemblance stops. The B-18 engine has a 5-mainbearing crankshaft, with bearings that are remarkably generous in size; it is strong enough to withstand far more than is being asked of it at present. The block is much roomier than before, and at the present bore size of 3.31 in. there is no crowding. Water completely surrounds each cylinder, and that minimizes thermal distortion. An interesting feature carried over from previous Volvo engines is thermosiphon cooling for the cylinder block. This gives a very rapid warm-up around the cylinders, and that reduces bore wear—which is heaviest when the cylinder-wall temperature is below the dew-point of the corrosive vapors generated in the combustion process.

The cylinder head is blessed with valves and porting that would do justice to a racing engine. All of the ports are separate, and the inlets have inserted rings that perfectly match the manifolding to the ports. The engine is equipped with a pair of SU carburetors. The compression ratio is only 8.5:1, but—oddly enough—at the specified spark setting, the engine would not run on regular-grade fuels without some pinging.

Prior to the change of engines, Volvo had redesigned the old 3-speed transmission into an all-synchro, 4-speed unit: a change that was much welcomed. However, the extra gear was crowded in at some expense in strength, and there were some instances where owner exuberance resulted in the need for repairs. Concurrently with the B-18 engine, Volvo designed and developed an all-new 4-speed transmission with a greater torque capacity and an absolutely unbeatable synchromesh on all forward gears. The gear lever, a long stalk growing up out of the transmission tunnel and inclined back to bring the knob within easy reach, is unchanged. It would be nice (and much appreciated by all of us here) if Volvo would use the transmission extension provided on the P-1800 to bring the lever mounting back nearer the driver, thereby shortening the lever itself, and reducing the "throw" required.

Only detail changes, and exceedingly minor ones at that, have been made in the 544's chassis since its introduction. The front wheels are carried on unequal-length A-arms, and a very light and precise cam-and-roller steering is used. The rear axle, which has hypoid-type gears, is located by trailing links and a transverse track rod. Coil springs and telescopic dampers are used all around.

All of the other Volvos have gone over to disc brakes at the front wheels, but the 544 retains 9-in. drum brakes. Consequently, the 544's braking performance is not as good as the others', but it is still quite good. Our braking tests produced a strong odor of scorched lining, but no perceptible fade.

One of the more attractive features of the 544 is its sturdy and rattle-free unit-constructed body. Window area is a bit limited, as the posts are quite thick, and the styling is neither contemporary nor classic-beautiful, but the use of heavy-gauge sheet steel, and a lot of it, renders the 544 nearly indestructible.

In the interest of making the passengers as bash-resistant as the car, Volvo has developed a seat-belt that is one of the best. It is a strap that starts on the floor, leads across the lap to a latch-fitting on the drive tunnel, then goes up and across the chest, and then back to an anchor on the window post.

Volvo: the only way to fly?

On the new 544, the instrumentation has been changed to bring it more into line with modern practice, and padding has been added along the top of the dash. The speedometer is now one of those creeping horizontal-line contrivances, and while it may look better than the previous round instrument, it is by no means as readable. The end of the thermometer line is cut on a sharp angle, and one never knows whether to read the point, middle or heel of the slanted end. In checking speedometer error, we used the middle; the error was moderate at that point.

The *circa*-1948 bodywork of the 544 makes for a rather

The Volvo's lines are dated, but not entirely unattractive.

ROAD TEST
VOLVO PV-544

SCALE: 10" DIVISIONS

DIMENSIONS

Wheelbase, in.........102.5
Tread, f and r......51.0/51.7
Over-all length, in......175.0
    width..............62.5
    height.............61.5
    equivalent vol, cu ft...390
Frontal area, sq ft......21.4
Ground clearance, in......7.5
Steering ratio, o/a......n.a.
    turns, lock to lock....3.2
    turning circle, ft......32
Hip room, front......2 x 20.7
Hip room, rear..........51.5
Pedal to seat back, max..40.0
Floor to ground..........11.7

CALCULATED DATA

Lb/hp (test wt).........27.8
Cu ft/ton mile..........82.1
Mph/1000 rpm (4th)......18.4
Engine revs/mile........3270
Piston travel, ft/mile....1720
Rpm @ 2500 ft/min......4760
    equivalent mph.........87
R&T wear index..........56.2

SPECIFICATIONS

List price............$2330
Curb weight, lb........2160
Test weight............2500
    distribution, %.....52/48
Tire size............5.90-15
Brake swept area........n.a.
Engine type......4-cyl, ohv
Bore & stroke....3.31 x 3.15
Displacement, cc.......1780
    cu in................108.5
Compression ratio........8.5
Bhp @ rpm........90 @ 5000
    equivalent mph..........92
Torque, lb-ft......105 @ 3500
    equivalent mph..........64

GEAR RATIOS

4th (1.00)..............4.10
3rd (1.36)..............5.57
2nd (1.99)..............8.16
1st (3.13).............12.8

SPEEDOMETER ERROR

30 mph..........actual, 29.0
60 mph................57.9

PERFORMANCE

Top speed (4th), mph......92
    Shifts, rpm-mph
    3rd (5500)............74
    2nd (5500)............51
    1st (5600)............33

FUEL CONSUMPTION

Normal range, mpg.....25-29

ACCELERATION

0-30 mph, sec..........4.3
0-40....................6.8
0-50....................9.6
0-60...................14.1
0-70...................19.1
0-80...................27.0
0-100
Standing ¼ mile.......19.1
    speed at end..........70

TAPLEY DATA

4th, maximum gradient, %..8.4
3rd....................12.6
2nd....................19.3
Total drag at 60 mph, lb..150

ENGINE SPEED IN GEARS

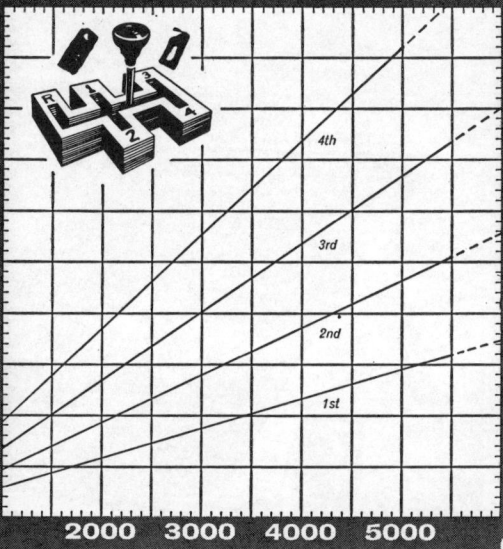

ACCELERATION & COASTING

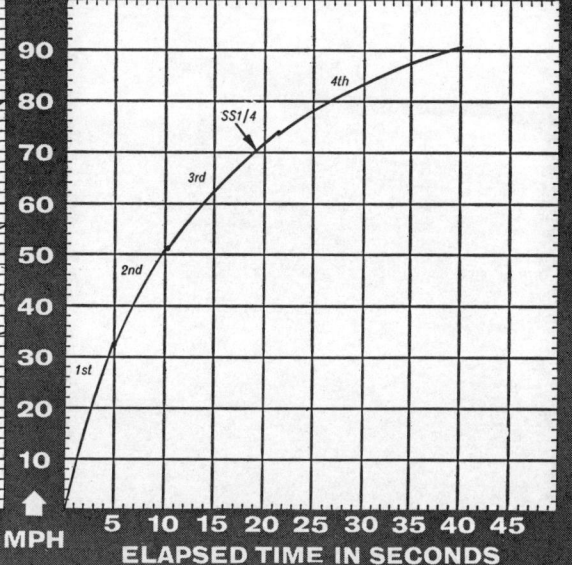

81

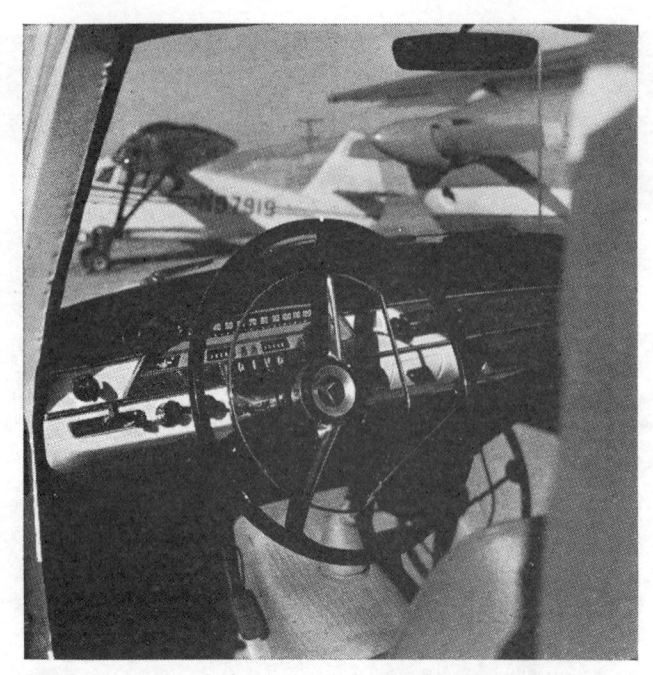

Nicely grouped and clearly labeled controls and instruments.

Accessibility is a requirement that has been fully met.

VOLVO PV-544

narrow interior, but there is adequate shoulder room, and a *lot* of head room. This is one of the few imports that one can drive while wearing a hat—if that matters. Leg room has been supplied unstintingly, but the area around the pedals is a trifle narrow for comfort. The seats are well contoured, and the placement of the controls, relative to the seats, makes this rather a nice car for long trips—much better, in fact, than many another car with nominally more posh interior. The upholstery is all done in a durable and rich-looking polyvinyl plastic, and there are a lot of nice small touches: such as an ash tray at each end of the back seat and back windows that pivot out for ventilation. Everything, except a radio, is included in the basic price of the car—and that includes a venti- lation and heating system that really does the job as it should.

Trunk room is good by import standards; fair as compared to most U.S.-built compacts: adequate, in any case, for the average family on the average trip (as any married man knows, there can never be *enough* space). At the other end of the car, room has been provided around the engine to make routine service less bother than is so often the case.

Above all, the Volvo 544 is a practical car. Its relatively light weight and small overall size, combined with what is really a very good chassis, make it a pleasure to drive, but its most valuable attributes are economy and durability. True, it cannot match the real midgets for mileage, but it does not have their lackluster performance or limited load capacity, either. If the Volvo has a single most-attractive feature, it is sturdiness and overall quality. There is nothing slap-dash or flimsy anywhere on the car, and this is, in our opinion, more than enough to compensate for any lack of sheer glamour.

Rather limited visibility astern.

Trunk space is well provided.

Often called "The Best Obsolete Car On The Road," it makes many owners happy. Why?

VOLVO 544

RETAIL: $2,395
WHOLESALE: $1,935

Among the curious facts turned up in ROAD TEST's continuing survey of car owners that an extraordinarily fierce loyalty exists in most Volvo drivers . . . a loyalty transcended only by that of Porsche pushers whose devotion amounts to a cult. To the non-owner the car reveals little which would account for such partisanship. The 544 is purely warmed-over Pre-War Ford in appearance. It is higher priced than most imports. Resale is an unknown quantity in most parts of the country. It is far from being the quietest thing on the road and performance is nothing to write a ream of copy about. Unlike the Porsche, it has no great competition story behind it nor any outstanding engineering advance at which to point with pride. "Fine Swedish Craftsmanship" can't be the whole answer, so what is it?

☐ Is the proud cry of the Volvo owner really a scream of pain filtered through teeth clenched in a grim smile? Is the paean of praise actually the false report of the guy who has been had and wants his friends to get stung too so he won't be alone?

☐ The Volve enigma has been a puzzle to the ROAD TEST staff for some time, so a rather more thorough research effort was decided upon . . . it to include an in-depth owner quiz and a comprehensive survey of service records. Since the same basic model has been in produc-

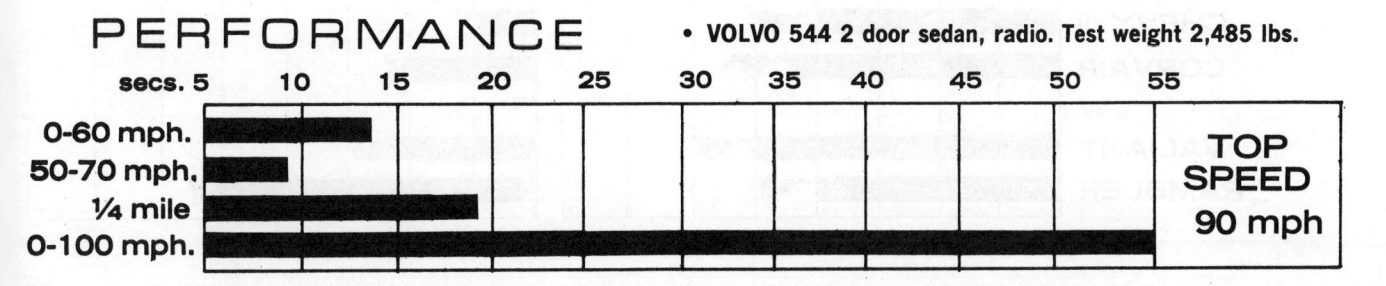

PERFORMANCE

• VOLVO 544 2 door sedan, radio. Test weight 2,485 lbs.

secs.	5	10	15	20	25	30	35	40	45	50	55	
0-60 mph.												TOP SPEED 90 mph
50-70 mph.												
¼ mile												
0-100 mph.												

83

HOW THEY COMPARE

PRICE

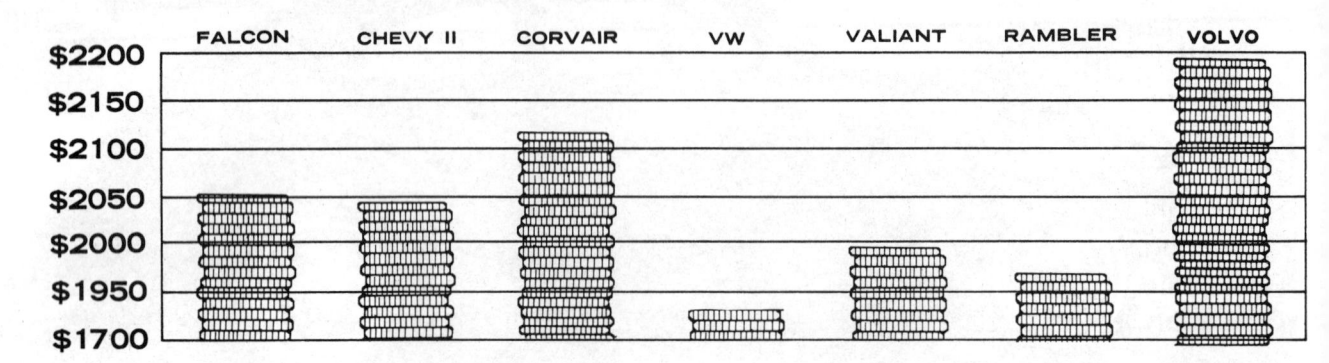

	FALCON	CHEVY II	CORVAIR	VW	VALIANT	RAMBLER	VOLVO
$2200							
$2150							
$2100			■				
$2050	■	■					■
$2000					■		
$1950						■	
$1700				■			

WHEELBASE

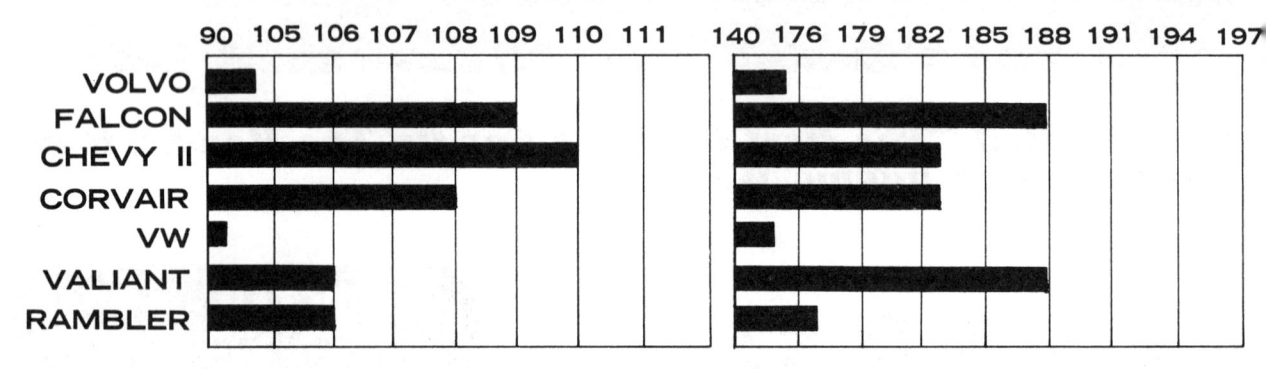

	90	105	106	107	108	109	110	111
VOLVO								
FALCON								
CHEVY II								
CORVAIR								
VW								
VALIANT								
RAMBLER								

OVERALL LENGTH

140 176 179 182 185 188 191 194 197

TURNING CIRCLE

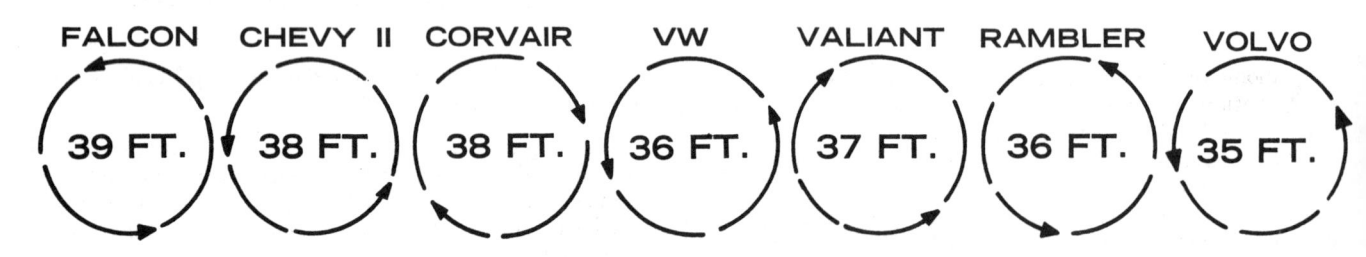

FALCON	CHEVY II	CORVAIR	VW	VALIANT	RAMBLER	VOLVO
39 FT.	38 FT.	38 FT.	36 FT.	37 FT.	36 FT.	35 FT.

HORSE POWER

40 50 60 70 80 90 100 110

VOLVO, FALCON, CHEVY II, CORVAIR, VW, VALIANT, RAMBLER

DISPLACEMENT

60 150 160 170 180 190 200 210 220

84

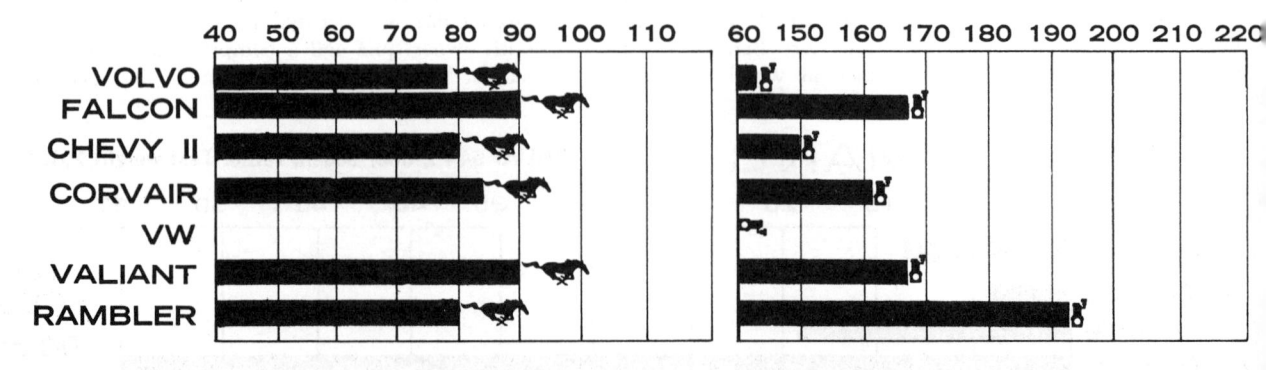

HOW THEY COMPARE

WEIGHT

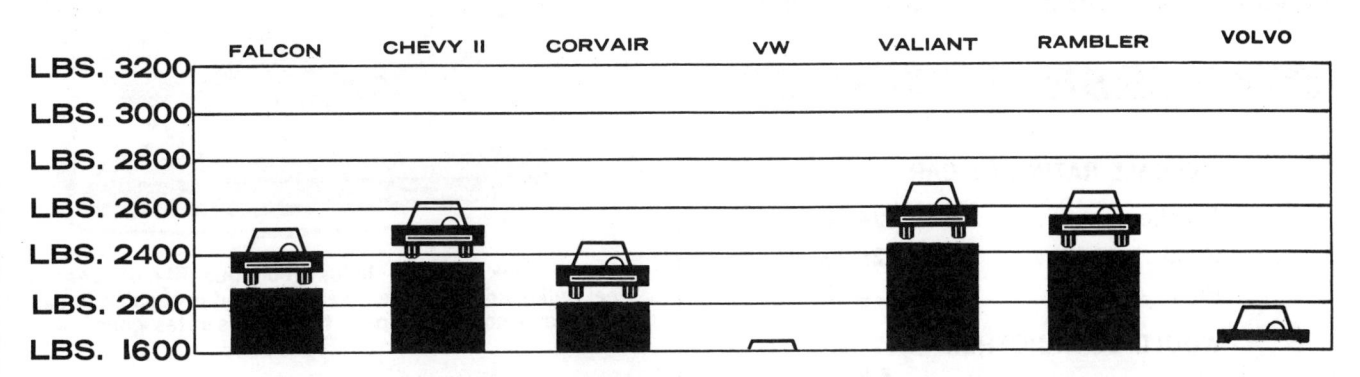

	FALCON	CHEVY II	CORVAIR	VW	VALIANT	RAMBLER	VOLVO
lbs.	~2250	~2400	~2300	~1650	~2500	~2500	~1750

BRAKE FACTOR

VOLVO	60
FALCON	52
CHEVY II	65
CORVAIR	68
VW	74
VALIANT	58
RAMBLER	55

TIRE FACTOR

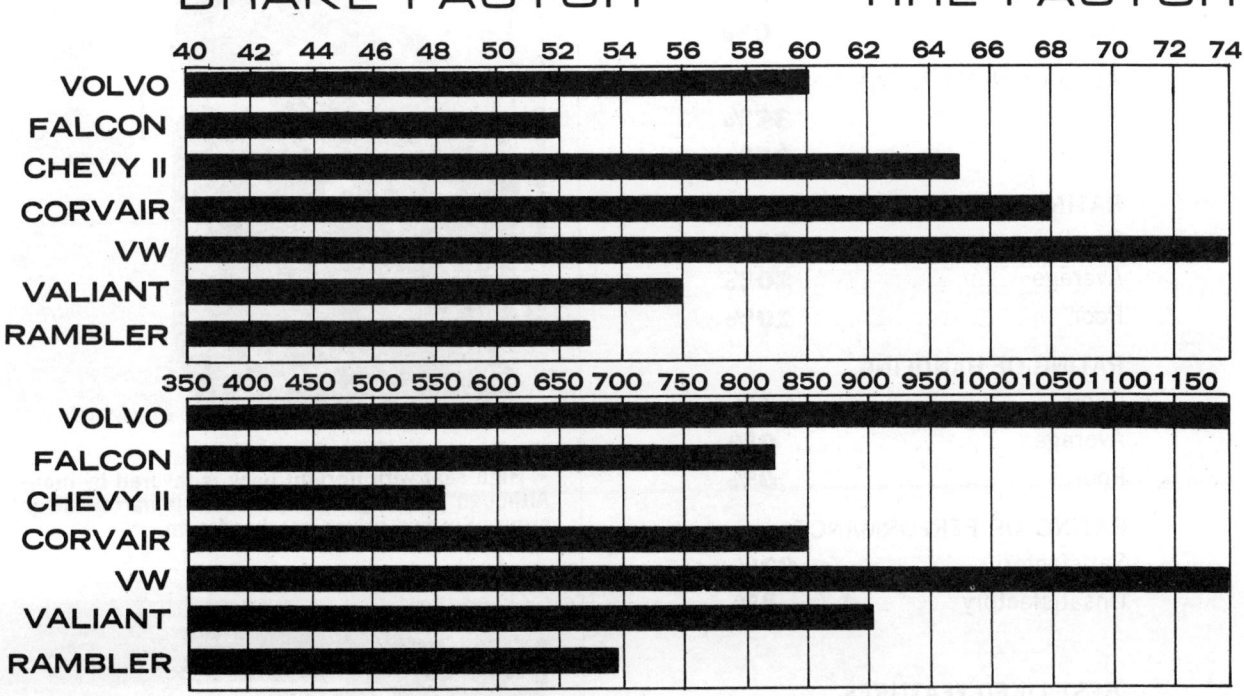

VOLVO	1150+
FALCON	850
CHEVY II	550
CORVAIR	850
VW	1150+
VALIANT	950
RAMBLER	500

ECONOMY FACTOR

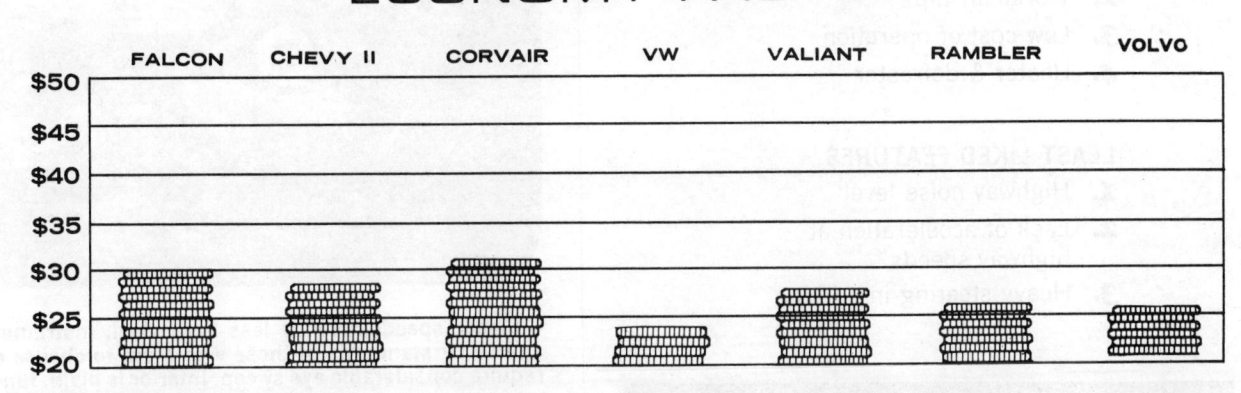

	FALCON	CHEVY II	CORVAIR	VW	VALIANT	RAMBLER	VOLVO
$	~30	~28	~31	~22	~27	~27	~25

WHAT IS A 544?

544 is the model designation of Volvo's low-priced car. It is available in only one body style, a 2-door sedan, with no optional engines or transmissions. Standard equipment includes: heater, shoulder harness seat belts, electric wipers, padded visors, cigar lighter, bucket seats.

OWNERSHIP SURVEY

OVERALL RATING OF CAR
Excellent **75%**
Good **20%**
Poor **5%**

WOULD BUY ANOTHER
Yes **63%**
Maybe **28%**
No **9%**

ALSO OWN AMERICAN CAR
Yes **35%**
No **65%**

RATING OF DEALER SERVICE
Good **70%**
Average **20%**
Poor **10%**

RATING OF HANDLING
Good **91%**
Average **9%**
Poor **0%**

RATING OF PERFORMANCE
Satisfactory **92%**
Unsatisfactory **8%**

BEST LIKED FEATURES
1. Handling
2. Workmanship
3. Low cost of operation
4. Heater & defroster

LEAST LIKED FEATURES
1. Highway noise level
2. Lack of acceleration at highway speeds
3. Heavy steering in parking

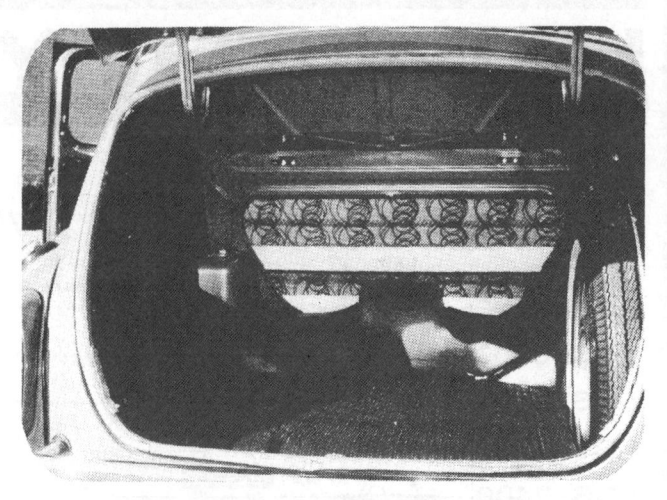

• Volvo trunk is deep, holds good quantity of luggage. Steel bulkhead behind rear seat is added protection in case of rear end collision. Spare tire access rates good.

• High seat with upright back is favored by many drivers. Although passenger compartment is narrow height is generous. Six two driver has headroom.

• Linear speedometer is less than ideal, instruments are sparse by standards of those who seem to choose car and require considerable eye sweep. Interior is plain, functional.

tion for many years, the findings can be brought forward with validity to apply to the 1965 model.

☐ The 544 is a two door sedan mounted on a 102½" wheelbase and powered by a four cylinder engine of 109 cu. in. developing 90 hp at 5000 rpm and 105 lbs. ft. of torque at 3500 rpm. Compression ratio is a modest 8.5 to 1, but the car requires premium fuel for best performance and freedom from "running-on." An excellent heater is standard equipment as are padded sunvisors and dash, cigar lighter, seat belts and electric wipers with washer. Standard extras not normally noted on a compact are a full horn ring and trip odometer. The wheels are also balanced at the factory and an a superior all-synchro 4 speed stick transmission is included in the base price of $2395.00 P.O.E. West Coast. The Volvo's size and price make it directly competitive with several domestic compacts, notably the Falcon and Valiant, but its image appeals more to the man whose family circumstances have caused him to leave two passenger sports cars than it does to the conventional compact buyer.

☐ In other words, the 544 Volvo is a car competitive with our low-priced vehicles, yet it seems to have a sporting flair or heritage which somehow sets it apart. (Volvos have competed with honor in road races and rallies for years and they will do so exactly as delivered. In fact, the ugly duckling 544 has been a lot more successful competitively than its beautiful GT sister, the P-1800 sports coupe.)

☐ Here is a rundown on how this car compares with the two American compacts mentioned. With comparable equipment, the six cylinder Falcon (170) costs about $150.00 less on the west coast. (But is not available with four speed transmission, which, if we use the accepted $185.00 to $200.00 option price as a guide, would make the Ford product cost slightly more than the Volvo.) For less money, then, the purchaser gets the extra smoothness of a six, but sacrifices several miles per gallon in economy. This doesn't amount to much in dollars, however, as the Falcon will use regular with aplomb. Although the Ford has 15 more hp, the overall performance of the cars is much the same, as the 170 six weighs enough more to make pounds per horsepower ratings almost identical. Due to softer spring rates and a longer wheelbase (by 7"), the Falcon can be expected to ride better on smooth roads, but the Volvo is much more stable.

☐ Compared to the Valiant series 100 which costs $2004.00 ex. factory, plus $186.00 for 4 speed transmission, $16.00 for padded dash, $17.00 for the wiper and washer group, $11.00 for bumper guards, $5.00 for smog control, $4.00 for cigar lighter and $17.00 for heavy-duty suspension, (a total of $2260.00 plus transportation to the west coast) the Volvo comes off less easily. The Valiant wheelbase is 3½ inches longer, it has eleven more horsepower and weighs nearly 700 lbs more. As delivered with the stock 101 hp engine, the Chrysler compact is no match in performance to the Volvo, due to the extra weight of the compact car, but it must be remembered that the optional 145 hp engine costs only $47.00 extra. With this additional power, the Valiant will run off and hide from the Swedish import, yet cost per mile is little different due to the fact that economy is still good and regular fuel is specified. With H. D. suspension, the Valiant handles every bit as well, but it suffers slightly in the braking department.

☐ Obviously, the Chevy II, the Rambler American, Studebaker and Corvair must also be compared, but our charts

• **Volvo B-18 engine now used in 544 has five main bearings, has better reliability than earlier type. Frequent tune-ups are required.**

will illustrate these without reference here.

ENGINEERING FEATURES:

■ Any engine with the ratio of 90 bhp to 109 cu. in. cannot be called a lightly stressed engine, but the Volvo, with its five main bearing crank, has a veriable reputation for reliability and major trouble free operation. The design is straight forward enough, much like the Chevy II four, in fact. Valves are pushrod operated in the conventional OHV configuration, but the extra power per cubic inch is developed by good breathing, a relatively wild cam and dual carburetion. The chassis and body feature up-to-date unit construction and suspension is by coil springs at all four corners, independent at the front; live axle located by torque arms at the rear. Brakes are conventional Bendix duoservo, but lining area is generous. Nothing about the car suggests any significant or breath-taking engineering advancements; it is simply a soundly engineered car put together with more than average care and skill — something like many American automobiles used to be.

CREATURE COMFORTS:

■ It has been said that the Swedes are a hardy folk, able to put up with discomforts the average American couldn't stomach. If this is so, the Volvo must be a truly deluxe vehicle on its home ground. As might be expected from a car bred in a sub-artic clime, the heater is first-rate and the defroster system works well enough to melt ice off a freezing windshield. People who live in cold climates will also appreciate the driver-controlled radiator blind, which enables extra heater efficiency and faster warm-up. The fresh-air ventilation system for summer operation has received less attention, however, and does not compare with that available in the cheapest domestic car. The electric wipers work well, as do the washers and it is commendable that they clean the normal blind spot adjacent to the driver's corner post. As previously mentioned, the visors are padded and swing to the side for side window glare. They are quite adequate in every respect. Less than adequate, however, is the interior mirror. This is not because of its design, but rather because the old-fashioned body includes a rear window which is very shallow. Even if the driver turns around in his seat, he is practically blind and cannot see the hood of a car at the rear while parking, let alone the rear of his own car.

☐ Vent wing controls are conventional latches and door and window handles are not inconvenient, although we have seen better. The seats, fortunately, have a broad range of adjustment to fit nearly everyone and the front buckets are shaped to hold driver and passenger fairly well during violent cornering. They are also of a comfortable shape, although rather hard for some tastes. However, the Volvo deserves no extra marks for interior storage space in the way many European cars do. The glove box is smallish and there are no storage spaces other than the dangerous one behind the rear seat and below the rear window. But arm rests are well placed; there is no glare from the top of the panel and all controls are convenient, well marked and within arms reach. Glare from the chrome horn ring can be annoying at certain times of the day, but otherwise the interior is quite satisfactory. The long, slim and rather flexible gearshift lever rather reminds one of an old Terraplane, but action is smooth and easy.

☐ Major impression of the interior of the 544 is that the manufacturer has done an excellent job of updating a very old design, incorporating many of the features desired by today's buyers, but struggling with parameters which cannot be changed. A case in point is the glare-free curved windshield surrounded by corner posts which are thick enough to hide a truck. These corner posts, coupled with rather small side windows and a miniscule rear window give a cooped-up feeling for which excellent interior design cannot quite compensate.

INSTRUMENTS & DRIVING CONTROLS:

■ The Volvo speedometer is of modern straight-line type and is well-placed for easy visibility, as are other instruments. The panel is only partially hidden by the horn ring and night lighting is good. Intensity may be driver controlled by a convenient rheostat. The speedometer proved to be of reasonable accuracy, as well, averaging only 6% fast. A fuel level gauge is included, as is a temperature gauge, but oil and generator indication are by warning lights. The trip odometer which can be reset is useful, but instrument calibration is sparse. Lack of complete instrumentation and complete calibration are unusual for imported cars in the Volvo category. We wonder why the manufacturer chose to save money here.

LUGGAGE ACCOMMODATION:

■ Although the trunk of the 544 certainly will not measure out as the biggest in the field, it certainly deserves top mark for being the most useful. The loading still is very low, the opening is large and the spare tire is tucked vertically at the side out of the way. It will handle four big suitcases with room left over for odds and ends and it is ideal for general hauling. It is also neatly trimmed and finished with a form fitting mat to protect luggage and the floor. A pair of skiis will not fit, however, and golfers might find the width too narrow for big bags, plus a cart or two.

DETAIL & FINISH:

■ When a manufacturer builds essentially the same product for 20 years, he should certainly learn how to de-bug a product and furthermore how to put it together with skill and care. It would be difficult to imagine any other car at the price which is as well assembled, detailed and trimmed as the Volvo. The paint is first class in appearance and lustre and is six coats deep. All doors, panels, weatherstripping, etc. fit properly and there are no rough edges. The chrome appears to be superior quality, as well it might be, to stand the rigors of North Country winters. The bumpers are solid and sturdy and are fitted with rugged over-riders. Gaudy ornamentation is conspicuous by its absence, with what little trim as exists being tastefully and solidly applied.

☐ Volvo doors close with a note of authority, although it is well to have a window cracked when closing them, as tight weatherstripping and sealing trap air pressure inside the car and prevent easy operation. Inside, the neat tailoring of the form-fitting rubber floor mats and the high quality feel of the upholstery is apparent. Here, too, everything fits properly and of all the Volvos we have inspected, not one looks as if it were assembled during a coffee break. All controls operate smoothly, windows work the way they should (quickly & easily) and pride in craftsmanship is evident everywhere.

TOWN & FREEWAY RIDE, HANDLING, PERFORMANCE & ECONOMY:

■ The Volvo 544 is a very easy and likeable car to drive when not pressed hard. The steering in particular deserves high praise. It is light, quick (3¼ turns to lock) and transmits little road shock . . . a good example of what good design, plus proper size and placement of the steering wheel can accomplish without power assist. Clutch and brake are light and positive in action and the throttle linkage is well coordinated. The car is lively and very responsive and, unlike many underpowered imports, it can hold its own under nearly every traffic condition. The gearbox, with its relatively close ratios is a joy to use and the pleasure is marred only by the wand-type lever.

☐ Fuel economy in city traffic was reported and verified at 18 to 20 miles per gallon when driven without regard for cost, although this can be improved by several mpg by conscious effort. At 65 mph, a high of 27.6 mpg was recorded, but an overall average is closer to 25 mpg. Cold weather seems to have no adverse affect on economy. Despite considerable use of the manual choke, mileage remains much the same.

☐ The ride of the 544 is a bit stiff-legged around town,

but it smooths out at speed. It handles dips quite well with no secondary tossing. The car feels like it will corner like a race car, but there are some oversteering tendencies until one becomes well accustomed to the fast steering. The ride is rated as acceptable and the handling as superior for a compact class vehicle. Brakes are excellent; pedal pressure is low and there is a good progressive feel. Road test crews encountered no fade at any normal speed, but high speed testing revealed the same faults experienced with self-energizing brakes on other automobiles. Such brakes are highly sensitive to water, dirt and other foreign matter and tend to become unpredictable when overheated.

TOURING & ROADABILITY

■ Generally speaking small imported cars born and bred in countries which are in most cases smaller in area than many of our states, are not well suited for cross-country touring in the American manner. Although the Volvo is intermediate in size, it also leaves a lot to be desired for such traveling. Speed is not the problem. It will cruise with reasonable lack of fuss at 75 or more and has a top speed in the low nineties. The car is somewhat wind sensitive, however, and the quick steering which is such a joy in town traffic and on winding back roads becomes too sensitive for relaxed driving at elevated speeds. This, combined with the smallish interior, wind noise and the constant drone of the carburetion and exhaust makes the car less than acceptable for this purpose. Thus it is not as smooth or as quiet as our compact sixes, but it does obtain somewhat better fuel economy at high speed than they do.

SUMMARY:

■ The average male driver will find the 544 handy and convenient much the way the 1942 Ford was. In fact, had the choice been available at that time he would probably have selected the Volvo instead. But the world moves on and the most charitable thing which can be said about the car in 1965 is that the maker has done an excellent job of updating an old-fashioned car and building it with pride and care. It is a rugged vehicle, but interviews with owners and service records reveal that it is not likely to cost less to repair and maintain over the years than one of our representative compacts. The timing gears have tended to become noisy at fairly early mileage, clutch judder is sometimes a problem and like most European cars, minor oil leaks are a constant bother. Its dual carburetion system requires careful synchronization and since the engine is in a relatively high stage of tune, it is particular about fuel and the frequency of tune-ups.

☐ We suspect that the major appeal of the 544 is to the young married man in the transient stage between a sports car and a full-sized automobile. He is still youthful enough that he does not want to give up the precise handling characteristics of his old open sports roadster, but wife and baby demand a roof over their heads and economy of operation is a factor. His loyalties to foreign cars have been stoked by years of reading ROAD & TRACK and, until the fires are banked, he will not be ready to view the automotive scene with an unprejudiced eye. In the meantime he will be driving a good automobile with average resale value, although if he needs to cash out in a hurry he may have difficulty finding a buyer with ready cash, particularly if he happens to live outside a metropolitan area.

☐ This customer, and others, would be well advised to consider the Volvo 122 series. This car sells for approximately $200 more than the 544 and offers a good deal more interior room, much better visibility and styling which is at least updated to the mid-fifties. As part of the bargain, disc brakes are included at the front, which by themselves are worth the extra money. The 122 rides slightly better due to a bit more weight and better distribution, it is considerably less wind-sensitive and quieter to ride in. Other than being susceptible to "freeway hop" in the same manner as the 544 (freeway hop is the pitching of the suspension in synchronization with tar strips on concrete highways at certain speeds), this deluxe model is much more acceptable for touring.

- **Handling of Volvo is rated as good by 75% of owners. Characteristics are those of stiffly-sprung pre-war Ford with fast steering. High speed cruising is somewhat unpleasant because of high noise level.**

Here is how "Jacob," the 1959 Volvo 544 appeared in the year it was new when Volvos were heralded as the "Family Sports Car." Later, Volvos gained the reputation of being eleven year cars.

Owner's report:
A VOLVO PV544
AFTER ELEVEN....

In 1961 with night rallies becoming popular, Jacob was fitted with a pair of driving lights, the installation being featured in Foreign Car Guide's March, 1961, issue.

Jacob's panel at the 27,000 mile point featured auxiliary engine gages, switches and pilot lights installed by his owner. Panel still looks modern today.

OWNER'S REPORT... A Volvo PV-544 After Eleven Years and 100,000 Miles by George N. Freund

As the Volvo legends go: "Old Volvos never die, but are sold to friends or relatives. Ninety percent of the Volvos sold in the United States ten years ago are still on the road today and Volvos last eleven years in Sweden." This account gives the history of an eleven-year-old Volvo which has been in photographs on the pages of Foreign Car Guide (World Car Guide's predecessor) for many of the past years.

"Jacob" was the name given to chassis number 212272 when it was purchased by the author on April 17, 1959. Since chassis number 212272 was this writer's first Volvo and his first brand new car as well, it seemed fitting that 212272 should bear the same name as did the first Volvo ever built.

In the early months of Jacob's life, he was used in daily commuting in New Jersey traffic for about forty miles per day. On weekends he was subjected to strenuous rallies, gymkhanas and drag races. Despite this weekend workout of high speed activity, Jacob was always ready to start the weekly routine on Monday mornings.

Although the schedule of rallies was rather extensive, truthfully though, Jacob was only drag-raced twice. In Jacob's first encounter at the one-quarter mile strip, he managed to defeat all comers except another Volvo like himself. The victorious Volvo was an old hand at racing, having brought his owner trophies on every weekend for the season. Jacob's second attack at the asphalt in April of 1961 brought him in a strong first place with a winning elapsed time of 18.55 seconds and a speed of 74 mph. It is worth noting that Jacob scored this feat with 28,000 miles on the odo while still using the original ignition points which remained until 70,000 miles. For comparison, a recent road test on the 1969 Volvo 164 recorded the ET for the quarter mile at 17.6. So old Jacob should be pretty proud of his record of nine years ago, especially since he cost only about half as much as his younger cousin does.

Deciding to remain triumphant at the strip, Jacob retired from racing and concentrated on the less demanding of motor sports, rallies and gymkhanas in which he brought home many trophies for his master, and in providing reliable daily transportation.

Reliable Jacob was. He never became ill away from home and started every time his owner demanded it. During Jacob's first 70,000 miles, he consumed fuel and oil at a very conservative rate. Jacob's average annual fuel consumption was 26.74 miles per gallon during his first year and was still 27.27 at the end of 70,000 miles. Oil consumption continued to decrease until Jacob had passed the 25,000 mile mark where consumption leveled off to a rate of 2,500 miles per quart where it remained until the odometer read over 85,000 miles. Today, at 101,000 miles, the original engine drives 1,500 miles before requiring a quart of oil.

Nearly all repairs and maintenance for Jacob were performed by his owner except specialized services like wheel alignment and balancing. Charges shown in the table reflect normal labor rates,

An appreciated feature in the Volvo 544 was a rear seat back that folded forward, allowing objects up to six feet in length to be carried with the trunk lid closed.

The B 16B engine was the one that brought Volvo racing success in 1959. Standard equipment included twin SUs, bright valve cover and windshield washers.

based on the owner's labor hours and rates which prevailed at the time.

In preparation for competition, Jacob's owner performed a valve grind at 7,830 miles. At this time, the valves were ground and re-seated, shims were installed to increase spring tension, the cylinder head was milled 0.030 inches and inlet and exhaust ports were ground to match the manifold ports. Normally, valve servicing would not be required at this low mileage and was not required on Jacob at the time. Perhaps this racing work aided Jacob's performance in his later years as now at 101,000 miles, compression is still high and uniform in all cylinders.

During a rally, the screws which secured Jacob's transmission front bearing retainer and oil seal became loose, allowing all the transmission oil to drain out which damaged the clutch friction disc and both mainshaft bearings in the transmission. This mishap occurred at 32,500 miles. When Jacob's transmission was reassembled, his master was sure to install the lock washers on the retainer nuts that the factory had chosen to forget when they built Jacob. The trouble never recurred. "Foreign Car Guide" featured Jacob in the January 1962 issue, showing how the transmission repair was carried out.

Jacob was not without a few annoyances, despite his general reliability. In warm weather, particularly in traffic, Jacob's SU carburetors would become over-heated allowing the gasoline to percolate, causing very rough idling and occasional stalling. A metal and asbestos heat shield was fitted between the float chambers and exhaust manifold. This shield relieved the hot-idle problems and a similar unit is now standard on all current model Volvos.

After about 30,000 miles, Jacob emitted an annoying grinding noise from the driveline at about fifty miles per hour. Thinking that hard driving had damaged the universal joints, Jacob's owner set out to replace them. Alas, after this effort the noise remained. Today, Jacob's owner recognizes the noise as common to most Volvos including current models so perhaps we shouldn't chastise Jacob for the expense of replacing his universal joints at such a young, tender age.

Jacob was easy on tires and retained the original Goodrich whitewalls until 38,000 miles despite much hard driving in rallies. Original brake linings lasted until 45,000 miles. The first set of replacement tires bought for Jacob were the expensive Dunlop SP variety which greatly improved Jacob's handling and traction on snow. Jacob never did require snow tires. Unfortunately, these special tires did not last as long as did the original equipment because cracks appeared in the side walls after two years. These tires were replaced after 32,000 miles with Firestone Deluxe Champions, which are still in use today.

Surely part of Jacob's good tire mileage can be attributed to the high quality king pin front suspension which has only required occasional alignment plus normal lubrication during the past ten years. Jacob is still able to pass the rigid New Jersey Motor Vehicle Inspection and all tires, though nearly bald, are all worn evenly. Free play in the king pins even now is barely noticeable, although lower control arm bushings are becoming loose.

In the fall of 1963, when Jacob had attained 69,000 miles, he fell to the tradition of remaining in the family. His original owner had ideas of owning a 122-S but couldn't bear to sell the treasured Jacob to a stranger. So Jacob changed hands and became the possession of the first owner's sister.

No more competition driving for Jacob now as he resigned himself to providing only reliable transportation for his school-teacher mistress. Before long, the school teacher became a housewife and Jacob became a second car for the first time, carting baby carriages, playpens and the like and taking his lady on many short hops to the store . . a far cry from his rally days. However, Jacob still remained true

Jacob today is not quite as shiny as he was eleven years ago but the rusty dents could easily be repaired.

to his mistress and only began showing signs of age after he reached the 90,000 mile mark when he needed a new starter, rebuilt carburetors, a regulator, universal joints and a new battery. His owner installed a 6-12 volt battery which provided the extra juice of 12 volts for starting that Jacob probably needed all his life, as the 6-volt system never quite matched the requirements for starting his high compression B16B engine.

Jacob was never a garage car and spent his entire lifetime braving the outdoors. However, the chrome on his bumpers still remains bright without a trace of rust, although the hub caps finally rusted through after ten years. Poor Jacob! Although his insides were well maintained with frequent oil changes and lubricatons, his exterior was subjected to a total of twelve crashes during the past ten years. Every time, except once, when Jacob was involved in an accident, he was standing still. So you see, Jacob was not at fault in any of his own disfigurements. Twice Jacob's brakes enabled him to stop quickly enough to prevent him from colliding with the car in front, but a power-braked equipped American car in back could not match Jacob's agility so rear bumpers were replaced twice. Several times Jacob was attacked by unknown cars while patiently waiting for his owner at the curb. Jacob was seen being repaired in a series of articles in FCG entitled "Bodywork for Beginners" appearing in April through August issues in 1965.

Today, Jacob is hardly as pretty as he was in April of 1959, since the most recent fender-benders remain as scars of age, showing rust where the once-proud paint chipped off under impact. Nevertheless, Jacob still drives as well as he did ten years ago. His upholstery of vinyl still looks like new, although the floor mat is now worn thin in a few spots. Jacob is still as spirited after 100,000 miles as he was in his youth.

Steering is still feather-light and the M4 gearbox still can be speed shifted without a single crunch, although the clutch does slip a bit on full-throttled shifts. Despite the weak clutch, Jacob manages to leap to sixty miles per hour in just under 12 seconds. His best time, long ago, was 10.5 seconds on the same stretch of road. Worn rubber bushings on the rear suspension allow the struts to emit a noticeable clanging noise when the accelerator is depressed and the extra-heavy-duty Gabriel shock absorbers that were replaced 50,000 miles ago still ride as hard as ever.

Although Jacob shares the driveway with a 1967 model 122S, he prefers to park beneath the shade of a tree and allows the youngster to park in the garage. Once again, with 103,000 miles on the clock, Jacob is being subjected to daily commuting, although he now travels a distance of 10 miles per day and is allowed to rest on weekends. Even in this type of traffic commuting, Jacob provides his owner with 22.9 miles per gallon and can still achieve 27 mpg on an occasional 60 mph trip. In fact, when his young 122-S cousin was back to the dealer's shop for warranty work, Jacob took his owner's family on their vacation.

Jacob has managed to live up to nearly every claim made for Volvos, even though we have more paved than unpaved roads in the United States. He proved his worth as the "family sports car," was never sold to a stranger but rather kept in the family, and is still in every day use providing economical operation to his owner after eleven years.

In order to put Jacob in tip-top shape for his twelfth year and second century on the odometer, all that would be required would be: New rubber front and rear suspension bushings, new tires and brake linings and possibly some body work, if he were to regain his original countenance.

Well done, good and faithful servant. Jacob, you are not for sale yet!

1959 VOLVO PV 544
TEN YEAR RECORD—REPAIR & REPLACEMENT

MILEAGE	ITEM	AMOUNT	NOTES
2,800	Rocker Adjustment Screw Broken	$ 2.60	
5,690	Parking light bulb and fuse	.55	
7,830	Grind and Seat Valves, Mill head	40.68	For racing-not routine
7,900	New Condenser (Mallory)	3.00	
8,015	Four Spark Plugs	4.00	
8,515	Heater Hose	2.75	
9,362	Balance and grind brake drums	8.00	
10,260	Headlamp	2.50	
11,924	Voltage Regulator	14.95	
17,980	Adjust Carb Floats and New Spark Plugs	7.00	
18,249	New Parking Light Assemblies (2)	10.40	Original rusted out.
22,900	Generator Brushes, Fan Belt, Adjust regulator	15.40	
25,400	Align Front End	10.00	
32,500	New Battery and Spark Plugs	39.00	
32,500	Rebuild Transmission, New Clutch plate	46.99	Oil leak damaged bearings (see FCG January 1962)
33,000	Wiper Blades	3.00	
34,422	Universal Joints	33.78	
36,451	Interior Lamp Bulb	.50	
36,469	Exhaust Pipe and Spark Plugs	22.75	
38,400	Tires (Dunlop SP)	201.90	
39,400	Balance Wheels	3.50	
42,300	Shock Absorbers—Front and Rear	62.00	
45,600	Reline Brakes	35.00	
48,100	Headlamp Bulb, Spark Plugs	5.85	
50,350	Heater and Radiator Hoses	5.71	
50,350	Fan Belt	1.95	
52,300	Align Front End	12.00	
55,650	Heater Repair	10.00	
60,100	Rebuild Speedometer and Spark Plugs	27.50	
62,700	Four Tires (Firestone Deluxe Champions)	108.00	
67,800	Exhaust System	35.00	
69,000	Headlamp Bulb	1.85	
71,500	Rubber Bushings in Rear Suspension	18.00	
71,800	New Points and Condenser and Plugs, Rotor and Cap Installed	19.65	
76,450	Universal Joints	35.42	
77,800	Speedometer Cable	5.60	
81,300	Rebuilt Carburetors, Spark Plugs	81.25	
84,650	Voltage Regulator	15.95	
89,700	Rebuilt Starter	44.50	
90,100	Battery	35.95	
93,600	Front Shock Absorbers, Spark Plugs	37.80	
95,680	Reline Brakes	37.50	
97,800	Universal Joints	36.58	
98,500	Clutch Linkage Repair	18.00	
99,000	Wiper Blades, Exhaust System	38.75	
101,000	Spark Plugs, Points, Condenser Installed	14.65	
	TOTAL	$1,217.71	

ROUTINE MAINTENANCE

75	Oil Changes @ 60¢ per qt. average cost	$ 158.00
75	Lubrications @ $2.00	150.00
20	Filter Changes @ $3.00	50.00
5	Wheel Bearing repackings	30.00
	Carburetor timing and miscellaneous adjustments (approximately)	50.00
	TOTAL	$ 448.00

TOTAL COST BREAKDOWN 1959 VOLVO PV 544

TEN YEARS—101,000 MILES

	TOTAL COST	COST PER MILE
VARIABLE COST		
Repair and Parts Replacement	$ 1,217.71	$ .0121
Maintenance	448.00	.0044
Gasoline (@25.3 MPG Avg., 35.9¢/Gal.)	1,460.00	.0145
	$ 3,125.71	$ .0310
FIXED COST		
Depreciation (Cost—$2,350; Present Value—$150)	$ 2,200.00	$ .0218
TOTAL	$ 5,325.71	$ 0.0528

SPECIFICATIONS
1959 VOLVO PV 544

ENGINE
- Type — 4-Cylinder, Overhead Valve—B16B
- Bore — 3.125 ins.
- Stroke — 3.15 ins.
- Displacement — 98 cubic inches (1.58 Liters)
- Compression Ratio — 8.2 to 1
- Horsepower — 85 @ 5,500 RPM
- Torque — 87 lbs ft. @ 3,500 RPM

TRANSMISSION — 4-Speed

REAR AXLE — Hypoid Type
- Ratio — 4.56 :1

WHEELBASE — 102½ ins.

LENGTH — 175 ins.

WIDTH — 62.75 ins.

CURB WEIGHT — 2,200 lbs.

STEERING
- Ratio — 15.5 to 1
- Turns—Lock to Lock — 3¼

(Continued from Page 63)

There seems to have been very little if any trouble developed in the field, but dealer service personnel have experienced cases where owners complain about overheating. Here it has been traced to failure to lower the radiator blind or a low water supply. Detonation or spark knock is usually caused by the use of lower than the recommended grade of 93-octane gas.

Other than occasional cases of neglect in lubricating the car (as directed in owner's manual lube charts) some cases of erratic running have been traced to failure to lubricate the carburetor or dampener at 1000-1500-mile intervals.

The vacuum piston in the carburetor dome is located and travels on a centrally located hollow spindle. An oil-controlled dampening piston is fitted in the spindle to prevent a rapid rise of the piston when the accelerator is floored. Check and maintain the oil level in the dampener spindle. Simply remove the brass hex nut on top of the dome's bottle neck, and withdraw nut and plunger. Maintain the oil level about ¾ way up the spindle. Experience has shown that S.A.E. 10 is better in winter and S.A.E. 20 in summer.

The fluid level in the brake master cylinder is another point that is sometimes neglected, and cases have been reported where the slip joint at the forward end of the rear half of the driveshaft has been overlooked at lubrication time. This lubrication point (see lube chart in owner's manual) is in addition to the lube fittings on the front and rear universal. It should be noted here that some universals have lube fittings and others are fitted with permanently lubricated joints that require no service. However, with either the lube or sealed universals the slip joint requires lubrication at 3000 mile intervals. A lubricant containing 80% graphite is recommended.

When the slip joint is neglected, it usually shows up as a ticking noise under sudden loading or acceleration. When badly in need of lubrication, the gear shift lever tends to move back obliquely to the left. This, of course, is caused by stiff movement of the driveshaft on the shaft splines.

A higher rated thermostat is available to improve heater performance in extreme cold weather areas. This optional unit starts to open in the 176-181-degree range and opens fully at 205 degrees, for those who want to really toast themselves. The standard 'stat is a 167-172-degree type that opens wide at 194 degrees.

Volvo's use of such things as Delco shock absorbers, Borg & Beck clutch, Spicer rear axle and Wagner self adjusting brakes (early Studebaker type) makes the car interesting from a parts replacement angle. The probability and degree of parts interchangeable is not known at this writing.

The distinctive Volvo PV outline survived virtually untouched from 1944 until 1965. This particular PV 444 dates from 1954

SUPER SWEDE

Some maintain the great Volvo success story began with the hunchback PV 444. Peter Nunn discovers the joys of this underrated 'classic'

The casual mention, in our July '82 club pages, of an article-to-come on the Volvo PV 444/544 series provoked an immediate and almost startling response. One enthusiast rang up virtually straightaway to tell us about his own PV 444, another wrote especially from Sweden to ask when the story was going to appear and several other readers made a point of commenting on the proposed feature, either directly or in passing.

Considering the cars were never officially made in right-hand-drive form, never imported into this country by Volvo during their combined 11-year production run and were, in retrospect, Sweden's answer to the humble Austin Cambridge (who said it must have been a daft question?), the degree of interest shown in the article is puzzling to say the least. But wait, there's more: according to Patrick Bunn, the helpful PV Registrar of the Volvo Owners Club, there are at the moment just 60-or-so PVs in the UK and not all of those are runners, in the proper sense of the term. Of the estimated dozen 444s in England, a mere five are thought to be roadworthy (only 20 PV chassis are registered with the VOC although Patrick Bunn says he has a good idea where most of the unregistered cars are) so where did this sudden burst of PV enthusiasm originate?

Creates much interest

Peter Lee, the youthful owner of the black PV 444 shown opposite and the enthusiast who telephoned us originally, provides one or two clues here. "Quite often I get people coming up to me in the street, asking about the car. One bloke, though, introduced himself one day and said: 'That's a nice car you've got there, how about doing a swap?' When I asked him what he had he replied: 'Oh I've got lots of cars you'd like, PA Crestas, Zodiacs, Jags – take your pick, any one you want'. Well of course I told him immediately I wasn't interested in that kind of thing, he could keep them all as far as I was concerned."

Having seen Peter's car at close quarters and had a brief turn behind the wheel, it's not difficult to see why it attracts so much attention. Quite simply, the PV 444 exudes character. It must also be one of the most charismatic cars Volvo have ever produced, standing head and shoulders above later models.

Even 39 years after its introduction (the first PV 444 was unveiled at the 1944 Stockholm Show), the car's styling still looks unusual, not to say dramatic, the hunched rear quarters, bulbous front wings, slatted grille and raked windscreen all serving to echo some of the more acceptable aspects of contemporary American styling. It could be argued Peugeot's 203 was another car that managed to do this successfully.

Today, the two cars continue to carry those appealing lines with considerable verve, the Volvo perhaps being able to turn more heads in the street than the Peugeot; but both, alas, are beginning to fall foul of homeland customisers and boy racers.

No fewer than 196,005 PV 444s were made in Gothenburg between 1944 and 1958. The PV 544 that followed, survived up to 1965, knocking up a respectable 243,995 production total in the process although sticklers for accuracy will point out the PV445 commercial/estate variant (confusingly, it later became known as the PV 210) outlived the latter car by being manufactured until 1969. Peter Lee's sound 'H' model PV 444, which hails from 1954/1955 is in fact a '54 car registered on January 1 1955.

The Lee family have strong links with Sweden, Peter, who is part Swedish, speaking the language fluently. Since they have, over the years, owned a long line of PVs there can't be much you can tell *them* about the breed. DBC 610 was picked up in Sweden last year (hence the unfamiliar registration plates) where it was offered for sale in a dealer's showroom.

The jazzy facia has strong American overtones

Indestructable: PV engines seem to last forever

Currently in regular use, it has a 544-derived B16A, 1583cc single carburettor engine which pushes out 60bhp as opposed to the original B4B's 44bhp.

The four-speed gearbox, similarly, comes from a 544 but Peter defends this lack of originality with the succinct comment: "I use the car everyday, in all kinds of weather. I don't want a concours car for that very reason – it's just too much of a hassle. The person who owned the car before me carried out the conversion quite successfully and I think, overall, he did a good job."

Despite its late powertrain, DBC 610 has many of the trimmings of the period that old car collectors seem to adore. The split front 'screen (and metal sun reflector above it), side-mounted spotlight, ridiculously slow vacuum wipers, jazzy dashboard, embossed 'Volvo' hubcaps, dangerous-looking bonnet mascot – all of them so evocative of the time the car was new. Incidentally, original bonnet mascots, along with other pieces of authentic chromework and interior trim are now fetching big money in Sweden. Radiator grilles and front overriders are a case in point; genuine '53 examples of the latter are, we hear, currently changing hands for a whopping £40 each while original grilles are even more sought after by PV aficionados.

Following my all-too-short drive in Peter Lee's PV I couldn't help thinking that from a sporting viewpoint, Volvo really lost their way when they came up with the 144 and 244. Whereas the old car has qualities that even today appeal to the enthusiast driver, the later 'tanks' – with the possible exception of the 244 GLT – most certainly do not. DBC 610, on the other hand, felt 'right' from the word go. It's probably been around the clock at least twice (Peter can't say for certain how many times!) yet 29 years on, it still powers away from lights in a very impressive fashion, the unburstable B16A engine delivering generous quantities of torque from seemingly any point in the rev range. You could probably thrash this car flat out all day with no ill effects whatsoever. Indeed Peter admits to opening it out on many an occasion "but nothing happens" he concedes, "it just winds itself up and goes on and on and on."

Beginning to show age

The car does show its age when it comes to changing gear, for movements via that long lever are languid and not a little sloppy. Synchromesh is fine, though, and the accompanying clutch action is light and precise. The steering? Well, that's a mite sloppy too (there was about 2ins freeplay when I tried the car) although Peter assured me that was about average! Brakes work well but need to be treated with respect. Inside the car is roomy, functional and, in places, delightfully ornate. It's also comfortable and very practical.

On the open road, the car gives the impression of being able to trample cheerfully over anything and everything that comes in its way. It is, without doubt, an extremely tough car and this general feeling of undestructability tends to prevail whether you're in command of the large steering wheel or sitting on one of the passenger seats. Sure-footed handling in either wet or dry, in conclusion, make driving this car a real pleasure, the PV inspiring confidence at all times.

PV prices are low at present. It should be possible to pick up reasonable examples of either 444 or 544 for well under £1000 although low-mileage, clean 544 Sports (the rally-based homologation specials) could command double that figure with ease. Cars needing work naturally come much cheaper. The keen PV clubs in Holland, West Germany and Sweden can help track down many obsolete parts as can Patrick Bunn of the Volvo OC. Unfortunately Patrick, who can be contacted at Flat 9, 11 Queensborough Terrace, London W2 3TB (tel: 01-727 2461) confirms that PVs, by and large, have yet to find their market in the UK, there being little apparent interest shown by the 'classic car' world as a whole or the dealer network. That's good news for buyers, of course, yet bad news for an enthusiast with a car to restore. Perhaps things will change.

■ **ROAD TEST:** VOLVO PV544 v SAAB 96

James Taylor puts two Swedish classics through their paces and comes away surprised at the results

SOLID SENSE

PHOTOS GARRY STUART

No-one I've asked has been able to give a satisfactory explanation of why Swedish-built private cars all had two doors in the fifties. If you wanted four doors in Sweden, it seems, you had to buy a taxi, an estate car, or an import. Whether by accident or design, the Swedes simply built their cars like that.

Both of Sweden's indigenous manufacturers had their successes, Volvo with the PV444 and Saab with the 92. The PV444 evolved into the PV544 in 1958, and the Saab 92 evolved into the 93 and then the 96 in 1960.

It was 1966 when Saab secured a German-built Ford V4 engine to give the 96 some more urge, but the old Volvo — known affectionately as the 'hunchback' because of its shape — had ceased production a year earlier.

On Swedish roads in the late-sixties and early-seventies, examples of both abounded, but now the Volvo is practically extinct except as an enthusiast's car, while the numbers of Saabs in everyday use is also on the decline.

These two models have something else in common: longevity. The PV444 was introduced in 1944 and stayed in production (latterly as the PV544) until 1965. That's a total of 21 years. The Saab lasted even longer, the first 92s appearing in 1950, the last 96 in 1980.

Swedes with style: Volvo PV544 and Saab 96 V4. Both were proved with illustrious competition histories

ROAD TEST: VOLVO PV544 v SAAB 96

HANDLING

THE basic layouts of these two cars are very different. The Saab has front-wheel-drive, with its engine positioned just ahead of the front wheels; the Volvo has traditional rear-wheel drive. Expectations, then, are also different. You anticipate a more modern feel to the Saab, and you get it.

The Saab is nicely balanced, and its handling reassuring. From the outside, body roll is very noticeable, but the supportive seats help to mask most of this body movement from the occupants.

Steering is heavy but precise, thanks to a rack-and-pinion system, and the car feels very well-behaved. Like the Volvo, it is built of heavy-gauge metal and is heavier than its appearance tends to suggest. This, together with good aerodynamics, means that the 96 keeps its line well on fast, straight stretches of road.

It is hard to conceive of one of these cars winning rallies — it seems so much like an ordinary family saloon. The only clue lies in the way the 96 feels as if it would plug on forever, and perhaps that was its main strength in rallies.

By contrast, the Volvo does feel like a rally car. It is very much heavier to drive, and this gives it an air of solidity which is less apparent in the Saab.

That weight also helps it to feel very stable on motorways, where the undertow from passing juggernauts fails to unsettle its composure. It ploughs on with a determined relentlessness which is undeniably impressive, and must have stood it in good stead when the going got tough in competition.

On bends, the weight and the rear-wheel drive combine to produce a fair amount of understeer, but the car stays firmly on the road and rarely gives its driver the slightest cause for concern.

Like the Saab, the Volvo rolls a lot on bends, but the occupants are much more aware of it because the seats are flat and relatively unsupportive.

PERFORMANCE

Front-wheel drive Saab has well balanced handling and ride

Heavy Volvo handles well, but prone to plenty of roll during cornering

WITH only 65bhp, the Saab was never designed to be a roadburner, but its character lies more in the manner of its going than in its outright performance. It is, briefly, a gutsy little car.

The Saab is quick enough, but not fast, although its sub-90mph maximum is good for a 1.5-litre car of this age, but not outstanding.

Most noticeable is the way in which the aerodynamic body shape helps to minimise wind noise (and fuel consumption), but this highly desirable quality once again fails to conjure up visions of a tough rally car scrabbling its way round the special stages.

It is also odd to find a column gearchange in a car with a competition pedigree, even though it works perfectly well and as quickly as you could wish.

The Saab must have been one of the very few cars to feature a freewheel by the time it went out of production.

This device was originally specified to mask some of the roughness of the early two-

The Saab's interior could present problems if the driver and front-seat passenger are not on good terms

PRACTICALITY

THE two-door design of both these cars is certainly a drawback if you intend to carry more than one passenger, although it does add to their sporting appearance.

Inside, while the Volvo wins on space, the Saab has much the better seats. Its interior is compact, though, and there could be problems if the driver and front seat passenger are not on good terms with one another!

As for the rear seat, it undoubtedly would take three abreast, but occupants would have to think very hard about where to put their legs.

The Saab's rear bench can be folded down and the rest of the interior arranged to form a bed. Alternatively, this fold-down feature permits long loads to be carried, which is a great practical bonus.

The fastback styling of both these cars makes their boots look small, but in fact the opposite is the case. Their rear seats are set well forward of the raked rear windows, leaving plenty of room for luggage below the deep rear parcel shelves.

One surprising deficiency in the Volvo is its heater. As Sweden is subject to some very low temperatures, you might expect a car designed and built there to excel in this department.

Although I was assured it was working properly, it failed to do much more than take the chill out of the passenger compartment on the particularly cold day we chose to take our photographs. By contrast, the interior of the Saab felt like a Turkish bath and proved a welcome haven in between cups of steaming hot coffee!

The interior trim of the Volvo is also surprisingly basic, with nothing more than rubber matting to cover the floor. The Saab does at least have carpets, though not of particularly high quality.

All Volvo PV544s are disadvantaged by being left-hand drive, and the thick B-pillar makes visibility at junctions difficult.

Spares for the Volvo are available from specialists without too much difficulty, but all have to be imported as the car was never sold in this country when new.

Enthusiasm for the PV

stroke engine, but it also made for trouble-free gearchanging and contributed to fuel economy, so Saab kept it for the later cars.

What it does is to allow the 96 to 'over-run' its engine and coast; normal drive is instantly restored by a whiff of throttle to match engine speed to road speed. It is simple to use, though a disadvantage is that it cannot be selected on the move.

Brakes in the Saab are servo-assisted and feel very modern, but in this respect the Volvo shows its age. Its unassisted brakes are heavy, though very positive, and protested with a loud rumble when required to haul the car up from speed on one occasion.

With 90bhp from its twin-carburettor B18 engine, the Volvo is also very much quicker than it looks: in its day, it was actually viewed as something of a performance saloon. Getting up to speed is no problem, because the floor change is positive and a delight to use, despite a formidably long lever.

The engine has plenty of low-down torque, and the Volvo pulls strongly in top gear even after being baulked on a hill. In fact, as I discovered, it is perfectly capable of pulling away from rest in third gear without protest.

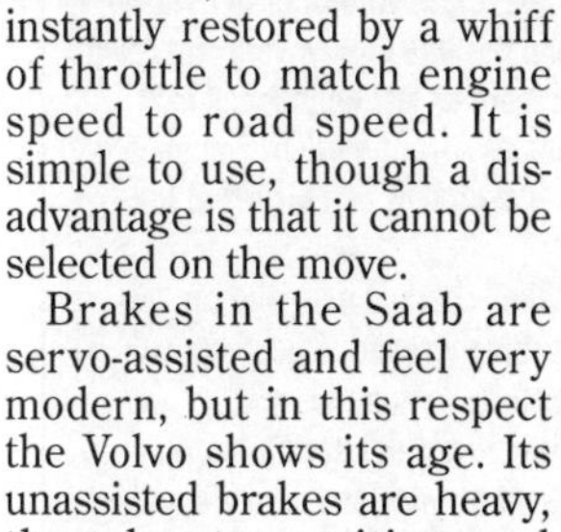

Saab V4 is 1492cc Ford unit: powers the front wheels

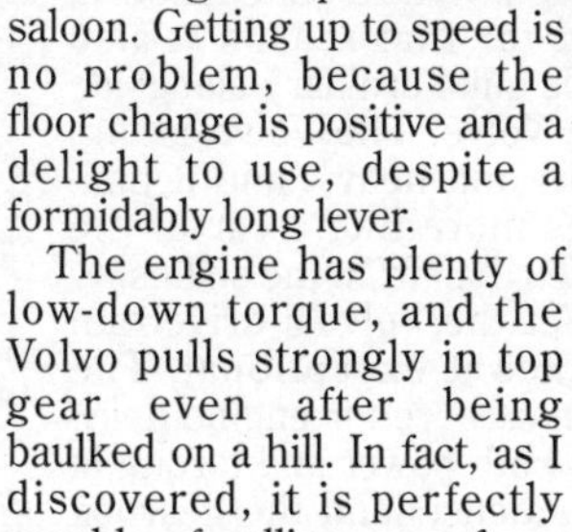

Twin carb Volvo B18 is lusty performer, though raucous

Column shift and flat floor characterise Saab, though seats are comfortable

Rubber mats on the Volvo's floor give a spartan image: facia similar to Amazon's

ROAD TEST: VOLVO PV544 v SAAB 96

models is more widespread overseas, but there is a growing following for the car in this country among people who want a Volvo that is different from an Amazon.

Most Saab spares are available more or less over the counter from Saab dealers. Scrapyards also contain cars aplenty to cannibalise, and there are several specialists who can help out, particularly through the thriving Saab Owners Club.

Saab's aeroplane ancestry shows in streamlined styling

VERDICT

BOTH these cars have attractive looks and distinctive personalities, and that makes either one of them an attractive choice for the enthusiast.

The Saab is a cosy little car with adequate performance and reasonable creature comforts.

Despite the age of its basic design, it feels quite modern, and for that reason it also feels quite ordinary alongside the Volvo. The Volvo is big, bluff and heavy, and it handles more like a car of the fifties than from the sixties.

Yet the Volvo's drivetrain makes it exceptional. The twin-carburettor engine delivers both power and torque in abundance, and the gearbox is a delight to use. It is more *fun* to drive than the Saab, and it leaves you wondering how modern Volvos could have become the safe and stodgy middle-class barges that they are.

To summarise: the Volvo is a real surprise, and definitely the one to have if you want unusual and distinctive styling.

But if you want practicality and modern creature comforts, you'd be well advised to go for the Saab. ∎

	Saab 96 V4	Volvo PV544 Sport
Year	1972	1965
Engine	V4-cylinder, ohv, single carburettor	4-cylinder, ohv, twin carburettor
Capacity	1498cc	1780cc
Bore/Stroke	90mm x 58.56mm	84.13mm x 80mm
Max power	65bhp @ 4700rpm	90bhp @ 5500rpm
Transmission	4-speed manual with freewheel	4-speed manual
Suspension	Front-wheel drive, with coil/wishbone ifs, dead tube rear axle	Coil/wishbone ifs with anti-roll bar, coil-sprung rigid rear axle
Brakes	Servo-assisted disc/drum	Drums all round
Steering	Rack and pinion	Cam and lever
Length	13ft 9.5in	14ft 9in
Width	5ft 2.5in	5ft 2.5in
Max speed	89mph	98mph
0-60mph	16.6sec	13.5sec
Fuel consumption	32mpg	28mpg

1965 VOLVO PV544 SPORT

THIS very smart car has covered no fewer than 127,000 miles, but was thoroughly overhauled during the winter of 1984-1985 by a former Volvo employee living in Sweden.

The work done was largely mechanical, and was designed to put the car into tip-top condition.

It reached this country last year through Volvo specialists, Classic Volvo Imports, of London (071 607 4025), and was snapped up by farmer, Chris Upjohn.

Chris' PV544 is representative of the penultimate 'F' series cars, built in 1964-1965. The final 'G' series of 1965 had an extra 5bhp squeezed out of their engines, but only 3400 were made.

The PV range was never sold in the UK, even though Volvo's imports of other models began as early as 1958.

1972 Saab 96 V4

SIMON Lucas has liked Saabs ever since his parents bought their first 95 — the estate variant of the 96. He found this one locally, very run down and more or less abandoned in someone's front garden, though it still had an MoT and a current tax disc.

When checked over, the car was surprisingly sound, a good enough reason for Simon to buy it.

The 96 did not need a huge amount of work, but second-hand wings were fitted all round by Simon, who also resprayed the car to put the car back on the road. Now the car needs some bright trim around the windows and a new rear seat.

The engine still pulls well, despite its 117,000 miles without major work of any kind, but Simon admits it is rather down on power and would benefit from a rebuild.

When that step comes, he will probably use the long-stroke crank of the 1700cc V4 engine — a simple conversion which improves torque and gives a little extra power.